HUMAN EVOLUTIONARY TREES

E. A. Thompson

Cambridge University Press

Cambridge

London · New York · Melbourne

CAMBRIDGE UNIVERSITY PRESS

Cambridge, New York, Melbourne, Madrid, Cape Town,
Singapore, São Paulo, Delhi, Tokyo, Mexico City

Cambridge University Press
The Edinburgh Building, Cambridge CB2 8RU, UK

Published in the United States of America by
Cambridge University Press, New York

www.cambridge.org
Information on this title: www.cambridge.org/9780521099455

First published 1975
Re-issued 2013

A catalogue record for this publication is available from the British Library

ISBN 978-0-521-09945-5 Paperback

Contents

Preface

This book is not a textbook of human population genetics, nor does it aim to provide general statistical methods. Its purpose is to present a detailed analysis of a specific problem concerning human evolution on the basis of a logically justifiable method of statistical inference. The problem is specific, yet methods of assessing the evolutionary relationships between populations (of the same or of different species) have attracted considerable interest since Charles Darwin first proposed the existence of such relationships. The method of inference is specific, yet it is one that must be at least an important facet in any complete scheme of scientific inference, and seems to be the only method which permits a unified approach to be taken to the analysis of data in the very wide variety of problems that arise in the field of population genetics.

The model through which inferences are to be made is also specific, and for this no apology is given. All scientific inference requires a model, and only when this model is explicit can the effect of its assumptions be investigated. Only by the analysis of data on the basis of explicit models appropriate to specific problems can hypotheses be objectively considered. In the case of population genetics problems, a model that can be fully analysed must probably always be a simplification of the true processes of evolution that have given rise to current genetic data. However, we must walk before we attempt to run: when the problems involved in the use of a simplified model have been solved, we may then proceed to extend the model in ways that will make it a closer approximation to reality.

Thus, although I believe the methods and results presented here to be of interest, and a detailed analysis of the particular problem to be of some practical importance, perhaps the most general aspect of the work is that of the line of approach. In the first chapter we place the problem in the more general field of inference problems in human population genetics, and consider previous approaches to it. We discuss also the view of inference to be taken in this work. Chapter 2 considers the genetic problem and its approximation by a probabilistic model. In Chapter 3 the mathematical analysis of the model is discussed, while Chapter 4 provides and investigates a method of making the required inferences. In Chapter 5 we consider the computational procedure and the estimates obtained for two particular sets of genetic data. Further problems and possible extensions of the model are also studied. In the final chapter an independent but related problem is investigated, and the approach is a repetition in miniature of Chapters 2 to 5: first the genetic problem, then the appropriate model, next the mathematical analysis of the model, and finally the analysis of some genetic data and a discussion of the results and of possible extensions of the model.

It is hoped that this book will be of interest to both geneticists and statisticians; it has not consciously been given either bias. Although some sections will be of greater interest to one rather than the other, it should be possible for the mathematics to be readily followed by the mathematically inclined geneticist, and the genetic discussion by the statistician with an interest in genetics. In the introduction of terminology and the provision of preliminary definitions I have intended to cater for both, but I have perhaps in general tended to assume the reader to have the same background as myself; that of a statistician whose interest in genetics, although not secondary, came later. Some knowledge of both subjects is necessarily assumed.

The majority of the research on which this book is based was carried out from 1971 to 1972 as a member of Newnham College, Cambridge, and as a research student in the Department of Pure Mathematics and Mathematical Statistics. The original research was supported by a Research Studentship from the Science Research Council, while latterly, during the writing of this book, I have been supported by a Sims Scholarship from the University of Cambridge. I am also grateful for the graduate scholarships and studentships I have held from Newnham College during this period. Chapters 2 to 5 are based on a research dissertation, awarded a Smith's Prize by the University of Cambridge (March 1973), while the material of Chapter 6 was first published by the Annals of

<u>Human Genetics</u> (37 (1973), 69-80). The work has more recently formed part of a thesis submitted for the Ph. D. degree in the University of Cambridge.

I am grateful to all those who have commented on or discussed any parts of this work. In particular I am indebted to Mr C. E. Thompson of the Computer Laboratory, Cambridge, for his advice on computer programming details and for other discussions, and to Dr J. Felsenstein of the University of Washington for the correspondence we have had on the subject of evolutionary trees. This correspondence raised several points of interest, and has contributed to the discussion presented in some parts of Chapter 5. Professor J. H. Edwards of Birmingham University provided the European genetic data on which the evolutionary tree of section 5.1 and the results of Chapter 6 are based. I am grateful also for a profitable week spent in his department.

Above all, I am indebted to my research supervisor, Dr A. W. F. Edwards of Gonville and Caius College, for his constant encouragement and for many helpful discussions. The extent to which this research has its foundations in his earlier work will become apparent, and I am grateful to him for the constructive interest he has taken in the progress of this work and in its publication. While it was through Dr Edwards that I first seriously encountered the problems of the foundation of inference and the subject of population genetics, I have greatly appreciated his encouragement of independent research and thought. The views expressed in this book are my own, as are, of course, any errors.

Cambridge
August 1974

E. A. Thompson

1· Inference and the evolutionary tree problem

1.1 PHYLOGENY, MODELS AND INFERENCE

The aim of this book is to provide a method of solution to a specific problem, and yet one which has attracted a wide interest in recent years. This is the problem of the statistical assessment of the phylogenetic relationships between various ethnic groups within the human species, on the basis of genetic data currently available in present-day populations. The basic difference between the approach to be considered here and that of some previous approaches is that inferences are to be based on a probabilistic model for the genetic evolution of the populations under consideration. The criteria of likelihood inference are to be used to assess alternative hypotheses of evolutionary history. No model can cover all aspects of the complex process of evolution, and inferences are necessarily made within the framework of the model. However statistical inferences cannot be made in the absence of a model, and, even if this model is necessarily a simplification of the true situation, an explicit statement of the assumptions under which inferences are made enables the effect of such assumptions and the possibilities of extending the model to be considered.

We shall consider only the problem of making inferences concerning several, often large, populations within the human species, these populations having a common source but having evolved largely independently, there being little interchange between them. Population differences reflect the length of time since the existence of a common ancestral population, and an evolutionary tree model is required. Some specific problems of population admixture may also be analysed on the basis of a model of independently evolving populations, and one such is considered in Chapter 6, but we shall not consider more generally the analysis of relationships between smaller populations where the pattern of differentiation depends mainly on the interchange between them and where migration has been sufficient for them to evolve substantially as a single unit.

Much work has been done on the mathematical analysis of migration models, and the genetic consequences of many specific migration patterns have been determined (see, amongst others, Kimura and Weiss (1964), Bodmer and Cavalli-Sforza (1968) and Maruyama (1972)). However, such models are complex and must involve many parameters if they are to bear any approximation to reality; problems of inferring migration history from currently available genetic data are largely unsolved except under equilibrium assumptions. Analyses of migration patterns (see, for example, Morton et al. (1968)) have been based on isolation by distance models (Wright (1943), Malecot (1959)), but the assumptions of uniform migration and equilibrium differentiation implicit in the model cannot normally be justifiable. Morton et al. (1971) have also developed methods for the study of genetic correlations between populations as measured by relative heterozygosities, but although these correlations provide measures of the patterns of population structure (Wright (1951)), they cannot be interpreted in terms of inferences concerning the history of populations in the absence of a model for this genetic history (Thompson (1974)). Thus the field of migration patterns is a further area in which likelihood analysis on the basis of explicit models appropriate to specific problems may perhaps provide an advance on present methods; but it is a field in which many problems remain to be solved, and is not one which we shall consider here.

A tree model does not allow for the existence of hybrid populations. While many populations are undoubtedly hybrid to some extent, substantial migration is a relatively recent phenomenon. In spite of the great increase in migration rates over the last few centuries, most individuals, even in the more mixed populations such as those of Brazil and Central America, may still be assigned at least a mixture of ethnic origins; most hybridisation is known. Thus the evolution of major populations may still be validly represented by a bifurcating tree; present genetic variation reflects the evolutionary history of populations for which a tree model is at least an adequate approximation. In the future this may no longer be so. At present migration rates relatively few generations must elapse before variation, even amongst some larger population groups, will

2

depend as much on recent migration patterns as on more remote evolutionary history. A reconstruction on the basis of a model of independent evolution would then no longer be a valid procedure.

The heuristic methods which have been previously used to provide phylogenetic representations of human populations are similar to those which have been used to estimate the evolutionary tree of the different species. In particular the method of minimum length spanning networks has been used by Dayhoff (1969) and that of least-squares additive trees by Fitch and Margoliash (1967) and Goodman et al. (1971) (see section 1.4). There has therefore been some tendency to consider the two problems equivalent. The obvious difference is the time scale. The evolutionary time, defined to be the length of time since the existence of a common ancestor, between man and his nearest neighbours on the tree of the species, the great apes, is at least 10^6 years and probably very much more. Homo sapiens evolved from Homo erectus only around $2\frac{1}{2} \times 10^5$ years ago (Cavalli-Sforza and Bodmer (1971: chapter 11)), and assuming a monophyletic origin the largest possible evolutionary time for any group of human populations is of the order of 10^4 generations.

There is however a far more fundamental difference. Differences between species are differences between the amino acid sequences of the various proteins which can occur. Differences between populations within a species are differences between the frequencies with which the various possible forms arise. In the former case the appropriate models are those of mutation in the discrete space of all possible amino acid sequences; in the latter the state space is the continuous space of the allele frequencies within each polymorphic system. The models which we shall consider are those of change of gene frequency, and not of change of gene state, and the methods to be developed are appropriate only to problems of frequency differentiation.

Still less are the methods to be regarded as general taxonomic procedures, although inferences are based upon population similarities and differences. In numerical taxonomy the aim is to obtain representations to the relationships between taxonomic units in the absence of any probabilistic model, sometimes in order to suggest hypotheses but more often simply as a classificatory or discriminatory procedure (Jardine and

Sibson (1971)). Our problem is to estimate certain parameters in a probability distribution derived from a specified model of evolution, and hence to judge between a priori specified hypotheses. The distance between populations is not a measure of taxonomic similarity but a random variable having an explicit distribution under the proposed model. This point has been made on several occasions (Cavalli-Sforza and Edwards (1967), Cavalli-Sforza and Bodmer (1971: p. 702)), but, because of the similarity of the heuristic approaches previously taken to this problem to some of the techniques of cluster analysis, the differences have not always been clearly stated. These methods are only justified by the belief that they provide an approximation to the estimate based on a full solution to the model, but while they are pursued with this view there is no justification for the criticisms of taxonomists (Jardine and Sibson (1971: p. 161)) that the solution obtained is a dendrogram which may not be interpreted as a phylogenetic tree.

In this section more emphasis has perhaps been laid on the problems which our analysis cannot be expected to solve than on those which it will, but a clarification of the assumptions of any model must always be of value. Morton et al. (1971) have criticised the use of tree models for the analysis of within species relationships. While it is true that no human population evolves in complete isolation, it is also the case that in many situations a tree model may be very much more appropriate than one of equilibrium differentiation under constant migration, and where there is a possibility of inferring the evolutionary history of such groups of populations it would seem to be a valid exercise to attempt to do so. All statistical inference requires a model, and, although the limitations of any model must be recognised, it is only through the analysis of data on the basis of models for which the problems of inference can be solved that progress will be made. When analyses are based upon explicit models we can consider the effect of the known assumptions upon possible results. When one problem is solved we can consider the possibility of extension to more general models which may be a closer approximation to reality in a wider variety of situations.

1.2 THE EVOLUTIONARY TREE PROBLEM

The model to be assumed for the evolution of human populations is one of an evolutionary tree. It is supposed that the populations under consideration are descended, by consecutive binary splitting, from some ancestral population existent less than 5×10^5 years ago. It is generally agreed amongst anthropologists that the evolution from Homo erectus (existent 5×10^5 years ago) to Homo sapiens (existent by 2×10^5 years ago) occurred only once, and thus that all populations may be assumed to be of monophyletic origin, although some would place the ancestral root at an earlier date. For most groups of populations it is unlikely that the effective ancestral population existed more than 5×10^4 years ago; only with expansion of numbers and movement of peoples will evolution on a tree model take place.

The data to be used in the statistical analysis of phylogenetic relationships are the allele frequencies at various blood group loci in present-day populations. It is assumed that during the process of evolution these gene frequencies have changed, independently at the separate loci, due to a process of random genetic drift (see section 2.1). By a series of transformations this process may be approximated by one of Brownian motion in a Euclidean space (see section 2.3). Thus a probability distribution for the present gene frequencies may be derived, given a 'history' of the populations consisting of the form of evolutionary tree, the times of split and the position of the initial root. The problem is to reconstruct this history from currently available genetic data (Fig. 1.2), or, in statistical terms, to estimate these parameters from observed random variables. A model for the splitting of populations may also be included. A simple birth, or Yule, process is the simplest appropriate model.

The original attempts to reconstruct the evolution of human populations from their sample gene frequencies used heuristic approaches (Edwards and Cavalli-Sforza (1963), Cavalli-Sforza and Edwards (1964)). It was realised that such approaches are insufficient in themselves and that any attempt to reconstruct evolution should be based on a probability model for the course of that evolution. The Brownian-Yule model was proposed by Cavalli-Sforza and Edwards (1967), but owing to difficulties

5

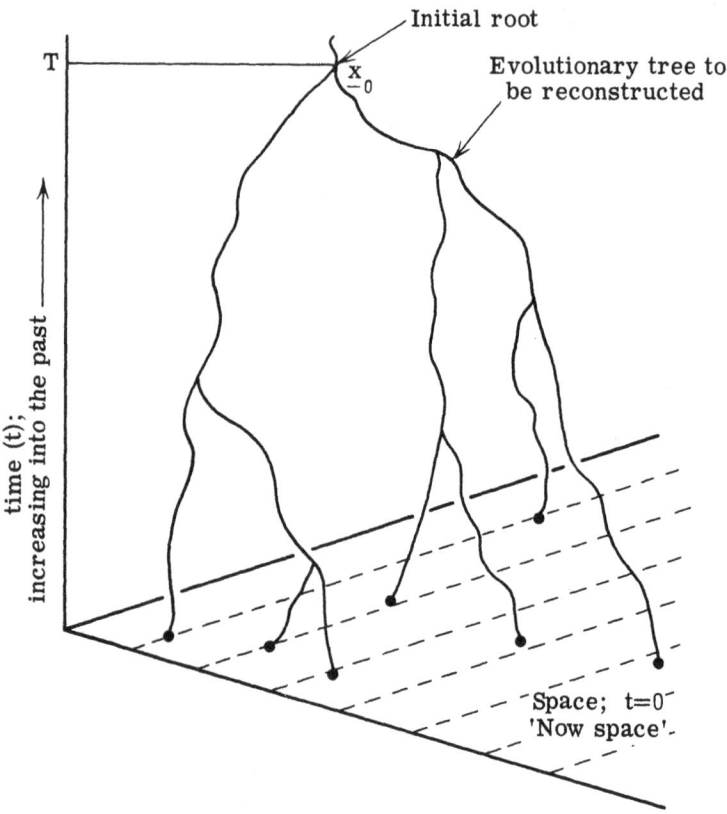

Fig. 1.2.   Representation of an evolutionary tree. T is the
total evolutionary time of this group of populations.
x_0 is the position of the initial root, or the point at
which the last common ancestor of this group of
populations splits. Information on population positions
(gene frequencies) is available in the space $t = 0$.

in analysis and 'singularities' caused by a confusion between the role of
parameters and random variables no analyses could be made on the basis
of the model. Writers continued to use minimum evolution (ME) and other
heuristic methods (see section 1.4) in the hope that these might provide
'a reasonable approximation to the likelihood solution' (Cavalli-Sforza
and Edwards (1967)).

The heuristic methods have been applied to several groups of populations (Ward and Neel (1970), Fitch and Neel (1969)). Although the populations often do not conform to the criterion of isolated non-inter-breeding units required by a tree model, the methods have proved successful at representing the relationships between populations in that the results obtained are compatible with known history and geographic and linguistic structure (Friedlaender et al. (1971)). However, such methods suffer deficiencies by comparison with maximum likelihood (ML) estimation of the evolutionary history (see section 5.6).

The model was improved as an approximation to reality by new genetic distance measures. Following an idea of Bhattacharyya (1946) for multinomial samples, Cavalli-Sforza and Edwards (1967) use a representation of the population gene frequencies at each locus on the surface of a multidimensional sphere. By stereographic projection of these spherical surfaces, Edwards (1971) obtains a Euclidean space of the required dimension, and populations may now be represented by points in this space. However, the model remained unanalysed; nor was the practical validity of these transformations investigated.

Progress was made by Gomberg (1966) in actually setting down the required probability distributions. Felsenstein (1968), dropping the Yule process and considering only the Brownian motion, developed a method of transforming the data variables in a way which simplifies the form of the likelihood and enables it to be evaluated, at given parameter values, from pairwise genetic distances. Edwards (1970) completed the first stage in the solution of the problem by giving the first fully correct statement of it.

Felsenstein (1973) has more recently used his transformation method to develop a method of rapid evaluation of the likelihood at given parameter values, and hence of maximum likelihood estimation of the evolutionary tree; a computer program has been written which searches for this maximum likelihood estimate by evaluation, for a given data set, at a series of parameter values. This approach is however essentially practical and, since it relies on the numerical evaluation of a single func-tion, gives little information about the shape of the likelihood surface, except for the given data in the region of the local maximum found. No full analysis of the model has previously been made; problems of existence and uniqueness of maxima have not been considered. Some assumptions implicit in previous work have not been justified, and fundamental points of likelihood theory remain to be fully clarified. The approach to be considered here takes Edwards (1970) as its starting point. Although an essential part is the development of a rapid iterative method of finding the ML tree, we emphasise far more an analytic treatment which provides an understanding of the process and the likelihood functions involved.

1.3 LIKELIHOOD INFERENCE

The evolutionary tree problem is to be studied from the approach advocated by the likelihood theory of inference. A full account of the basic theory with reference to questions of scientific inference is given by Edwards (1972). We give here a summary of those points which particularly influence the analysis of our particular problem; many of the problems of inference which arise in Chapters 3 to 5 are relevant to current thinking in the field of likelihood inference.

Likelihood theory advocates that hypotheses should be judged on the basis of their likelihoods. The likelihood of a point hypothesis, H, given experimental data, D, is

$$L_D(H) = P(D|H), \qquad (1.3.1)$$

the probability of the observed data under the hypothesis. When only one set of data is under consideration the subscript, D, may be dropped. All the information in data D on the relative merits of two hypotheses is contained in the likelihood ratio,

$$L_D(H_1)/L_D(H_2), \qquad (1.3.2)$$

or in the support difference, $S_D(H_1) - S_D(H_2)$, where the support $S_D(H_i)$ is defined to be

$$\log_e[L_D(H_i)]. \qquad (1.3.3)$$

Support is determined only up to an additive constant; only support differences, between alternative hypotheses for the same data, have any sig-

8

nificance. In likelihood theory it is assumed that there is always some, explicit or implicit, alternative hypothesis.

Thus for simple point hypotheses H_1 and H_2 the only problem is to decide what support difference should lead to the rejection of one hypothesis in favour of the other. The level 2, or a likelihood ratio of e^2, has been suggested, and will be adopted here, but the support scale has no interpretation in terms of probability or other measure, and the interpretation of the support surface lies with the individual investigator (Edwards (1972: p. 33)). In a statistical analysis it is better to state the actual support difference between alternative hypotheses of interest than to fix a universal 'significance' level.

Suppose now that H_i is a composite hypothesis concerning a multi-dimensional parameter θ in some space Ω; say H_i is $\theta \in G_i \subset \Omega$. The likelihood ratio for H_1 against H_2 is defined as

$$\max_{\theta \in G_1} [L_D(\theta)] / \max_{\theta \in G_2} [L_D(\theta)] = \max_{\theta \in G_1} [P(D|\theta)] / \max_{\theta \in G_2} [P(D|\theta)],$$

and the support difference as

$$\max_{\theta \in G_1} [S_D(\theta)] - \max_{\theta \in G_2} [S_D(\theta)]. \tag{1.3.4}$$

For composite hypotheses the problem of degrees of freedom arises. If G_1 has greater dimensionality than G_2 then it is plausible that

$$\max_{\theta \in G_1} [S_D(\theta)] > \max_{\theta \in G_2} [S_D(\theta)]$$

even if there is no difference between the two hypotheses as explanations of the data. Classical likelihood-ratio significance-testing theory solves this problem by considering the asymptotic chi-squared distribution of the support difference, but it remains an unresolved problem in likelihood theory. Although simpler hypotheses are to be preferred, deterministic hypotheses $(P(D|H) = 1)$ are normally to be rejected on grounds of prior knowledge (Edwards (1972: p. 199)). Within the class of hypotheses specified by the probability model there is no intrinsic reason, within a finite set of given data, to give a 'bonus' to a hypothesis having some arbitrary set of restrictions however intuitive these may be. In comparing forms of evolutionary tree we find that the spaces G_i are of the same dimension and that the problem does not arise (Chapter 4). However when we wish to compare the information on phylogenetic relationships contained in different data sets, or to consider hypotheses of simultaneous splitting, the problem must be considered (Chapter 5).

The function $\max_{\theta \in G} [L(\theta)]$ is known as the maximum relative likelihood (MRL) with respect to G (Kalbfleisch and Sprott (1970)). One way of dealing with the problem of nuisance parameters is by maximising over them and considering the MRL. This may be misleading if there are many such parameters, and alternative methods of eliminating them have been developed by Barndorff-Nielsen (1971), using concepts of ancillarity. These conditioning methods have the serious disadvantage that, if a parameter is discarded as a nuisance parameter but it is later decided that it should be considered, a complete reanalysis is necessary. This may give different estimates to the parameters already considered. Use of the MRL ensures equality of joint and marginal estimates. This problem arises in evolutionary tree theory with reference to the root, $\underline{x}_0$, which may be considered as a nuisance parameter in estimating the tree (Felsenstein (1973)), but which we cometimes wish to consider (Chapter 5). This parameter may be eliminated using M-ancillarity (5.5), but normally no parameter will be completely disregarded, and if not all are required at any stage we consider the MRL. We note that all statements are thus, implicitly if not explicitly, simultaneous joint statements about all parameters.

A further problem is that of 'prediction'. Fisher (1956: p. 126) proposes a predictive likelihood for future random variables whose distribution depends on parameters which are unknown but on which there is information through the observed data. However, there is no clear consensus as to whether statements of probability or likelihood are appropriate, or whether any single such statement can be made. An equivalent problem arises in the theory of evolutionary trees. We shall not wish to make future predictions, but we shall wish to make inferences about unobserved random variables. The proposed solution (3.2) corresponds to Fisher's predictive likelihood.

The basis of likelihood theory is that information about a parameter cannot be expressed as a probability distribution, unless it arises as the result of a random procedure having a probability model. If this is not the case knowledge or beliefs must be expressed in terms of the point function, likelihood, and not a set function. The support function is invariant under one-to-one (1-1) transformations of the parameters, and support is additive over independent data sets. The area under a support, or likelihood, surface has no meaning. Any prior beliefs regarding hypotheses may be expressed by a 'prior likelihood' but not by a Bayesian prior distribution (Edwards (1972: p. 36)). Prior and experimental supports add to give posterior support. An uninformative experiment, or no prior information, is expressed by a constant (without loss of generality (w. l. o. g.) zero) support function.

Fisher, in a comment on Jeffreys (1938), states that 'likelihood must play in inductive reasoning a part analogous to that of probability in deductive problems' and should perhaps be accepted as a 'primitive postulate' rather than justified by repeated sampling arguments. Thus in the likelihood theory of inference, as opposed to the procedure of maximum likelihood estimation, we are not concerned with asymptotic properties or repeated sampling justifications, but with the complete support surface as a representation of the relative ability of a given set of point hypotheses to explain a given set of data. In theory we should examine the contours of the complete support surface, which may be multi-modal or even have singularities. For a large number of parameters this is impossible, and often only the maximum and the curvature at the maximum are considered. If the support function is quadratic these determine the complete surface. From the curvature at the maximum two-unit support limits (parameter values at which the support is two units less than at the maximum) may be determined on the assumption of a quadratic surface in the neighbourhood of the maximum. These are the likelihood equivalent of classical confidence limits. The adequacy of this procedure clearly depends on the properties of the complete surface, which should be considered wherever possible.

All likelihood inference is conditional on the model accepted for the data: testing the model falls outside its scope. This does not mean that the validity of the model is unimportant - indeed the success of likelihood inference in scientific problems must depend critically on the scientific reality of the model adopted. The model is part of the prior information and can only be rejected if some alternative is contemplated. There is no universal measure of the weight to be attached to a model: this must depend on the individual scientist (Fisher (1956: p. 21)).

Robustness thus plays no part in likelihood theory. We have no test statistics whose distributions under deviations from the model may be examined. The change in likelihood function under deviations from the assumed probability distribution is precisely that deviation. The change in ML estimate may be examined, but, as a statistic, this estimate has little significance in likelihood theory. There are two reasons why a probability model may be adopted and these are well illustrated by the Brownian-Yule model of 1.2. The Brownian motion is a description of a real physical process that is taking place and the parameters have an existence independent of the model. In such cases the validity lies solely in the accuracy of the approximations involved (to be discussed in Chapter 2). The weight attached to the model will depend on the belief in the physical process rather than on the data.

Alternatively a model may simply be a convenient summary of the data. This is the case with regard to the Yule process for the formation of populations. The weight attached to such a model will depend entirely on how convenient a summary of the data it proves itself to be, and, in contrast to a 'real process' model, it has no intrinsic weight. A 'significance test' showing that the model is not an adequate summary of the data will lead to its rejection. However, even for a 'real process' model significance tests, whether of support (Edwards (1972: p. 180)) or classical form, may be of some importance in influencing our belief that the assumed process is the one that is taking place (2.4).

1.4 THE HEURISTIC METHODS

There are three main methods that have been used to reconstruct evolutionary trees from gene frequency data. These are 'minimum evolution', 'least-squares additive trees' and the method of Malyutov et al. (1972). These are summarised here for completeness, and so that their results may be compared with the likelihood solution to the problem. The interpretation of a representation obtained by these techniques as a phylogenetic tree is justified only by the assumption that the result is an approximation to the estimate of the evolutionary tree on the basis of some probabilistic model for evolution; this belief should therefore be investigated. In minimum evolution, for example, there is no assumption that evolution procedes in any minimal way (Cavalli-Sforza and Edwards (1967)).

(i) Minimum Evolution (ME)

This method was proposed by Edwards and Cavalli-Sforza (1963). It has been extensively used and produces acceptable trees (1.2). For this reason comparisons between ME and the likelihood solution, to be made in sections 3.3, 5.1 and 5.6, are of some importance. The aim of the method is to construct the minimum length spanning network, or minimal Steiner tree, between n population points, when these are embedded in a Euclidean space of (n - 1) dimensions, in accordance with the pairwise distances given by some genetic distance measure satisfying Euclidean metric conditions. The minimal Steiner tree is unrooted; no scale of, even relative, time is inferred. Thus, at best, the ME solution is a projection of the evolutionary tree of Fig. 1.2 into the current 'now space', $t = 0$.

Cavalli-Sforza and Edwards (1967 and other papers) have determined suitable algorithms for the construction of minimal Steiner trees. Since there are $\Pi_{r=1}^{n-2}(2r - 1)$ unrooted labelled tree forms, the major problem is to find a good initial tree at which to start iteration for the solution. The basis of the algorithms at present in use is described by Thompson (1973a) where a new method of finding an initial tree is suggested. The methods and programs, originally due to Edwards and Corfield (Edwards (1966)) but modified for greater efficiency, now seem to be in their most efficient possible form, unless and until a direct solution to the Steiner problem is found (Thompson (1973a)).

(ii) Least-squares additive trees (LSA)

The LSA method was suggested by Cavalli-Sforza and Edwards (1964) and details of the solution are given by Cavalli-Sforza and Edwards

(1967). The pairwise distances between populations are fitted, according to a least squares criterion, by additive distances along the internal branches of a given form of spanning network. The tree form having smallest residual sum of squares amongst those having positive estimates of the lengths of internal branches is to be adopted, but, as for ME, it can be positively identified only by examining all tree forms.

The tree produced is again unrooted, and although ME and LSA give very similar results LSA seems to have even less justification than ME as an approximation to the likelihood solution. The proposed model of gene frequency differentiation is one in which the population points move in a Euclidean space as the populations evolve in time, and we shall find that differentiation due to random genetic drift implies that mean square distances are additive over independent branches of evolution. The LSA method assumes that simple distances are additive, and it is not necessarily possible to embed the LSA solution in any Euclidean space. The method has been investigated by Kidd and Sgaramella-Zonta (1971) and other criteria for the adoption of a tree form have been suggested. LSA has been less used than ME in problems of gene frequency variation, but it has been considered extensively with reference to the problem of reconstructing the tree of the different species from data on protein amino acid sequences.

(iii) The Malyutov and Rychkov method

This method of backwards reconstruction is described by Malyutov et al. (1972) and is an advance on previous methods in that it is based on the probability model and produces a rooted tree with time scale and estimates of times of split. A tree form is predetermined by some cluster analysis criterion. Then joins are made, proceeding backwards into the past, estimating the time and position of each ancestor on the basis only of the population sizes and distances in time and space between the two populations to be joined. These two populations are then discarded and only the position and time of their new-found common ancestor are considered when the next join is made. Although the method is very rapid, the fact that the estimated position and time of any ancestral population depends only on those of its two immediate descendants, and not on those of its ancestors, results in very unreasonable trees. In practice the

14

solutions obtained are far worse than those provided by ME and LSA, particularly in many dimensions.

Even if the estimation criteria were those of likelihood, estimates based only on the two populations to be joined cannot approximate any overall method of estimation based on a complete model for all populations. The method does have the advantage that known differing population sizes may be taken into account, but more often these are not known for ancestral populations and different guesses may lead to widely differing results. Further the method provides no criterion by which the trees obtained may be judged, or by which the estimates resulting from different a priori assumed forms may be objectively compared.

Thus the tree methods that have been used in practice have major practical and theoretical limitations. There is a need for a practically viable and theoretically justifiable solution. Felsenstein (1973) has developed such a method of assessing any proposed evolutionary tree by evaluation of the likelihood; we shall provide a likelihood analysis of the model which enables the adequacy of locally-maximum likelihood estimates to be investigated in terms of the overall support surface, and provides a greater understanding of the interrelation between the observed data and the estimated tree.

2·The model

2.1 RANDOM GENETIC DRIFT AND THE PROBABILITY MODEL

Our aim is the 'reconstruction of an evolutionary tree', but it is necessary to define more precisely what is to be inferred from the data. Edwards (1970) defines types of tree as follows: a tree form is a tree specified only by its topology (as, for example, by Harding (1971)); a labelled tree form is a tree form where now the tips of the branches, the present populations, are distinguished. For n populations there are $\prod_{r=1}^{(n-1)}(2r - 1)$ labelled rooted tree forms (Cavalli-Sforza and Edwards (1967)). If further a distinction is made between trees of the same labelled tree form having different orderings in time of the internal nodes, we have a labelled history. It is the labelled history of the populations that is to be inferred, and the labelled history will in future be referred to simply as the form of the tree. There are

$$n! (n - 1)! / 2^{n-1} \qquad (2.1.1)$$

labelled histories for n populations (Edwards (1970)).

It is assumed that the major cause of gene frequency differentiation in contemporary populations is random genetic drift in preceding generations. [Random genetic drift (r. g. d.) is the name given to changes in gene frequency due to finite population size.] This model of neutral isoalleles will not be correct for all gene loci; for many characteristics gene frequency differences may reflect mainly differences in environment. The widespread polymorphism of the human blood group loci is discussed by Cavalli-Sforza and Bodmer (1971: pp. 732-8): many models have been proposed to explain how such a high degree of genetic variability could arise and be maintained. However, Kimura and Ohta (1971: chapter 9) have shown, on the basis of the theory of neutral isoalleles, that high levels of polymorphism may be maintained without stabilising selection.

For the majority of the human blood groups the observed levels of polymorphism and patterns of differentiation could well be the result of r. g. d. alone; whereas selection may or may not occur, r. g. d. must.

R. g. d. may be formulated as follows. In each generation, at any k-allele locus, the genes in a diploid population with effective size N_e may be considered to be a random multinomial sample of the $2N_e$ genes of the previous one. Thus if the gene frequencies at generation t are $\underline{p}^{(t)} = (p_1^{(t)}, \ldots, p_k^{(t)})$,

$$\sum_{i=1}^k p_i^{(t)} = 1, \text{ and } \underline{p}^{(t+1)} = \underline{p}^{(t)} + \underline{\varepsilon}^{(t)},$$

then $E(\underline{\varepsilon}^{(t)}) = \underline{0}$, $\text{var}(\varepsilon_i^{(t)}) = p_i^{(t)}(1 - p_i^{(t)})/2N_e$

and $\text{cov}(\varepsilon_i^{(t)}, \varepsilon_j^{(t)}) = -p_i^{(t)}p_j^{(t)}/2N_e$.

$$\left. \right\} \qquad (2.1.2)$$

N_e is the 'variance effective population size', which is a modification of the census size N taking into account non-random mating, age structure and geographic structure of the population. It is defined to be the number such that (2.1.2) is a correct description of the drift variances; for human populations prior to the very recent increase in longevity it is often estimated that N_e is of the order of $\frac{1}{2}N$.

Now the generations are not in reality discrete and (2.1.2) may be transformed to a process in continuous time. Let $\underline{z}(t)$ be the gene frequencies at time t, time being measured in generations, and let

$$\underline{z}(t + \delta t) = \underline{z}(t) + \underline{\varepsilon}(\underline{z}; \delta t).$$

Then $E(\varepsilon_i(\underline{z}; \delta t)) = 0$ for $i = 1, \ldots, k$,

$E((\varepsilon_i(\underline{z}; \delta t))^2) = z_i(1 - z_i)\delta t/2N_e + o(\delta t)$,

$E(\varepsilon_i(\underline{z}; \delta t)\varepsilon_j(\underline{z}; \delta t)) = -z_i z_j \delta t/2N_e + o(\delta t)$, $[i \neq j]$,

and all higher moments are of order $o(\delta t)$. Also (see, for example, the method of Kimura (1964)) the Kolmogorov forward equation giving the probability density $f(\underline{z}; t)$ of gene frequencies $\underline{z}$ at time t may be written

$$\frac{\delta f(\underline{z};\, t)}{\delta t} = -\sum_{i=1}^{k} \frac{\delta}{\delta z_i}[M_i(\underline{z})\, f(\underline{z};\, t)] + \tfrac{1}{2}\sum_{i=1}^{k}\sum_{j=1}^{k}\frac{\delta^2}{\delta z_i \delta z_j}\,[V_{ij}(\underline{z})f(\underline{z};\, t)],$$

(2.1.3)

where

$$M_i(\underline{z}) = \lim_{\delta t \to 0}\,[E(\varepsilon_i(\underline{z};\, \delta t)/\delta t)]$$

and

$$V_{ij}(\underline{z}) = \lim_{\delta t \to 0}\,[E(\varepsilon_i(\underline{z};\, \delta t)\,\varepsilon_j(\underline{z};\, \delta t)/\delta t)].$$

Thus for the case of r. g. d. we have

$$\frac{\delta f(\underline{z};\, t)}{\delta t} = (1/4N_e)[\sum_{i=1}^{k}\frac{\delta^2}{\delta z_i^2}[z_i(1-z_i)f(\underline{z};\, t)] - \sum_{\substack{1\le i\le k\\ 1\le j\le k\\ i\ne j}}\frac{\delta^2}{\delta z_i \delta z_j}\,[z_i z_j f(\underline{z};t)]],$$

(2.1.4)

with $0 \le z_i \le 1$, $\sum_{i=1}^{k} z_i = 1$ and $f(\underline{z};\, 0) = \delta(\underline{z} - \underline{p})$, the Dirac δ-function, for diffusion from some initial gene frequencies $\underline{p}$ $[\underline{z}(0) = \underline{p}]$.

We note that (2.1.3) and (2.1.4) are not the standard forms of the diffusion equations, but are completely equivalent to them. For

$$E([\sum_{i=1}^{k}\varepsilon_i(\underline{z};\, \delta t)]^2) = 0,$$

and under (2.1.4), or more generally under any model of frequency differentiation, the diffusion takes place with probability one in the space

$$\sum_{i=1}^{k} z_i(t) = \sum_{i=1}^{k} p_i = 1.$$

Thus if we write

$$g(\underline{z}^*;\, t) = g((z_1,\, \ldots,\, z_{k-1});\, t) = f((z_1,\, \ldots,\, z_{k-1},\, 1 - \sum_{i=1}^{k} z_i);\, t),$$

and similarly consider V_{ij}, M_i and z_k as functions of $\underline{z}^*$, we obtain the more usual diffusion equation given by Kimura (1964);

$$\frac{\delta g(\underline{z}^*;t)}{\delta t} = (1/4N_e)[\sum_{i=1}^{k-1}\frac{\delta^2}{\delta z_i^2}[z_i(1-z_i)g(\underline{z}^*;t)] - 2\sum_{\substack{i,\, j\le (k-1)\\ i<j}}\frac{\delta^2}{\delta z_i \delta z_j}[z_i z_j g(\underline{z}^*;t)]],$$

(2.1.5)

18

with $0 \le z_i \le 1$, $\sum_{i=1}^{k-1} z_i \le 1$ and the distribution at time 0 again being concentrated entirely at $\underline{p}$. (2.1.4) may be considered to contain redundent information, but its greater symmetry makes it preferable when transformations of $\underline{z}$ are to be considered.

Ewens (1965) has investigated, and confirmed for sufficiently large N_e, the validity of the transition from discrete to continuous time. The variance $V_{ij}(\underline{z})$ is sufficient at the boundaries $(z_i = 0 \text{ or } 1)$ to ensure that these belong to the domain of the diffusion, which is therefore closed.

The model assumed for population splitting is that these evolve independently, and in any time interval δt each population has probability $\lambda \delta t$ of splitting into two independent populations, each with the gene frequencies of the parent. That is, we have a Yule process. It is assumed that there is a 'last common ancestor', the most recent ancestral population from which all those under consideration are descended, and that this ancestor existed and split at most 10^4 generations ago. The gene frequencies of this ancestral population are a basic parameter of the model, as also is the population size N_e, assumed equal for all populations (but see also 5.3).

2.2 THE GENETIC AND HISTORICAL ADEQUACY OF THE MODEL

The validity of the model depends largely on the populations and gene loci to which it is applied. The first requirement is that we should choose populations for which there are sufficient data: it is important that samples should be sufficiently large for the apparent differentiation between populations due to sampling to be negligible compared with the true differentiation due to drift. Since sampling is, like r.g.d. itself, a multinomial sampling, this is a requirement that

$$(1/m) << (T/N_e) \qquad \text{(see 5.3)} \qquad (2.2.1)$$

where m is the number of individuals sampled and T the total evolutionary time of the populations in generations (see Fig. 1.2). Thus, for example, for the major populations of Western Europe we have perhaps $N_e \approx 10^6$, $T \approx 200$, and $m >> 5,000$ is required. In practice there are rarely data from sufficiently large samples, except for the ABO and Rhesus systems, and time estimates may be inflated due to sampling errors. Although a labelled history (2.1) may be correctly inferred, estimates of the times of split must be treated with caution: the sampling problem is discussed further in section 5.3.

The model does not allow for the existence of selection, migration or hybridisation. The units of population should be such that there is sufficient internal migration for them to be regarded as a single gene pool but very little migration between them. Many populations fulfil this requirement, at least until very recent generations, and although there are sometimes significantly differing gene frequencies between units within a population differences are substantially less than those between larger units, which are maintained by geographic, cultural, linguistic and political boundaries.

A small amount of migration is tolerable provided differentiation between populations is maintained. It may then be regarded in the same light as mutation - a factor scarcely affecting polymorphic gene frequencies, but a source of new alleles and an indication of the mutant/migrant nature of their possessors. The use of known migrants should be avoided in the population samples used: we require the gene frequencies for the descendants of some specified ancestral population. This is the opposite view to that for a migration model, where the study is essentially that of hybrids and migrants and the correlation of genetic and geographic distance. In practice evolutionarily close populations are often also geographically close. In these cases migration will tend to lead to underestimates of the evolutionary times. We may mention here 'migrated units' of population (Jews, American Negroes, Caucasian Australians, etc.), which may be treated as separate populations, and are distinct from migration in the above sense of mixing of populations. The ancestry of these units may be inferred from their gene frequencies, although in such cases there may be environmental selection effects. These will be indistinguishable from the effects of local admixture, both tending to make the migrated unit more similar to the local population, although selection will affect only some loci (Cavalli-Sforza and Bodmer (1971: p. 495)).

20

Thirdly we consider the problem of environmental selection. A selection model can never be refuted since by postulating the required selection coefficients any pattern of gene frequency variation may be obtained, and the coefficients involved would be too small to be detected from sample data. Although there is little confirmed evidence of selection in blood groups, there may be some effects on resistance to certain infections and environmental conditions. The problem is really one of the choice of suitable selectively neutral gene loci (see below) rather than of populations, but it has been suggested that only populations of similar environment should be used (Malyutov et al. (1972)). However, it is not known precisely what selection effects are to be avoided and hence in what ways the environments should be similar. There is however another reason for choosing populations that are not too widely separated: this is the problem of gene fixation. The Brownian motion dimension depends on the number of alleles, and if some allele is lost from a population the dimension is decreased. Thus we require populations that are polymorphic for the same alleles at the same gene loci. This is more likely to be true of populations not too widely evolutionarily separated.

It will be shown (5. 3) that constancy of populations size in time is not necessary for the validity of the model. However, for simplicity we shall assume it until that point. At any one time the population sizes should all be equal. This is a severe restriction, but it will be more nearly met if only populations of the same type are considered; that is, villages, tribes, countries or continents but not some of each. The model assumes instantaneous splitting and subpopulation sizes both equal to that of the parent. However, population splitting will generally take place at times of rapid population expansion and the parent population size will be rapidly attained. Moreover, although splitting will seldom be instantaneous, there will come a point at which migration will no longer maintain similarity and this may be said to be the splitting point. Migration subsequent to the split will decrease time estimates: non-random splitting of populations will increase them.

Thus, regarding population formation, there are many factors that may result in unreliable time estimates, but none of these should seriously affect the possibility of inferring a correct tree form. It should not be expected that consistent estimates of evolutionary times, or even relative times, will be obtained using different groups of populations of varying size and degree of isolation. However, although the maximum likelihood estimate of the evolutionary tree must be interpreted with caution, the likelihood surface as a whole provides the relative degree of support for any alternative hypotheses of evolution which may be expressed in tree form. Such a hypothesis is a description of major evolutionary events; it does not entail a detailed belief in the instantaneous occurrence of bifurcating splits at specific points in time followed by complete isolation of populations.

In choosing gene loci on which to base a phylogenetic study, we require unlinked polymorphisms for which there are sufficient accurate data and which are not subject to selection, particularly environmental selection. Constant directional selection will act equally on all populations, and so should cause less distortion of the pattern of gene frequency differentiation (Cavalli-Sforza and Edwards (1967)). In practice we use the red and white blood cell groups; none of those used show evidence of linkage. Under selection gene frequencies change linearly in time, whereas for r. g. d. it is the square of the change that is proportional to time. Thus the avoidance of selection is particularly important where large times are involved. Little is known about blood group selection, but the length of time for which polymorphisms have existed shows that directional selection is unlikely to be a major factor in frequency differentiation; for those loci used there is no confirmed evidence of widespread environmental or stabilising selection. The exclusive use of blood group loci has been criticised on the grounds that these are not a random sample of gene loci, but they are used because they are precisely those loci which may be expected to conform to the model. For a taxonomic procedure, where the aim is one of efficient classification, the use of anthropometric characteristics would clearly be more effective. In a phylogenetic study the aim is to reconstruct history according to a probability model and there would be nothing to be gained by considering loci that do not satisfy that model.

As will be shown (2.3), in order for the Brownian motion approximations to hold we must have allele frequencies that are not too small

22

and evolutionary times that are not too large. In practice only alleles with frequencies of at least 3% and preferably 5% in all populations should be considered separately, but low frequency alleles may be considered as a single class. J. H. Edwards (personal communication) suggests the use of only the most frequent allele at each locus, but this eliminates classes unnecessarily and useful information may be lost. The alleles must be grouped in the same way for all populations; each must have the same diffusion space. Provided r. g. d. is the differentiating force, this grouping does not invalidate the model in any way.

Cavalli-Sforza and Bodmer (1971: chapter 11) discuss the monophyletic evolution of Homo sapiens from Homo errectus and the subsequent evolution of human populations. They consider the formation of races by genetic isolation, their classification according to phylogeny, and the use of genetic polymorphisms as opposed to anthropometric characteristics for this purpose. Much of their discussion provides further justification for the approximation of human evolution by the proposed model of a bifurcating tree and r. g. d. , at least over that period of history that is relevant to current differentiation between major populations.

2.3 THE BROWNIAN MOTION APPROXIMATIONS

We have the diffusion equation (2.1.4) given by the diffusion means and variances;

$$
\left.
\begin{array}{l}
M_i(\underline{z}) = 0 \\
V_{ii}(\underline{z}) = z_i(1-z_i)/2N_e \\
\text{and} \quad V_{ij}(\underline{z}) = -z_iz_j/2N_e \quad (i \neq j)
\end{array}
\right\}
\left.
\begin{array}{l}
1 \leq i \leq k \\
1 \leq j \leq k.
\end{array}
\right\}
\quad (2.3.1)
$$

Cavalli-Sforza and Edwards (1967) note that the angular transformation $\theta_i = \cos^{-1}(z_i^{\frac{1}{2}})$ $(i = 1, \ldots, k)$ standardises the diffusion variance. The populations may then be represented by points on the surface of the unit k-dimensional sphere, $\underline{\theta}$ being the vector of direction cosines of the population point. The angular distance between populations with gene frequency vectors $\underline{z}^{(1)}$ and $\underline{z}^{(2)}$ at this k-allele locus is

$$
\phi = \cos^{-1}[\sum_{i=1}^{k} (z_i^{(1)}z_i^{(2)})^{\frac{1}{2}}].
\quad (2.3.2)
$$

The chord distance

$$[2(1 - \cos \phi)]^{\frac{1}{2}} \tag{2.3.3}$$

was suggested as an appropriate genetic distance measure, and it is now this distance that is most widely used in the various heuristic methods of reconstructing evolutionary trees.

Edwards (1971) shows the validity of the stabilisation of variance in the spherical space, and deduces the approximation of the process by Brownian motion on a sphere: however, not only the variance but the complete diffusion equation should be considered.

Transforming (2.1.3) the diffusion equation for θ becomes

$$\frac{\delta f(\theta; t)}{\delta t} = - \sum_{i=1}^{k} \frac{\delta}{\delta \theta_i} [M_i(\underline{\theta}) f(\underline{\theta}; t)] + \frac{1}{2} \sum_{i=1}^{k} \sum_{j=1}^{k} \frac{\delta^2}{\delta \theta_i \delta \theta_j} [V_{ij}(\underline{\theta}) f(\underline{\theta}; t)] \tag{2.3.4}$$

where $M_i(\underline{\theta})$, $V_{ij}(\underline{\theta})$ are defined in the same way as before (2.1).
Now

$$z_i = \cos^2(\theta_i)$$

and thus $\delta \theta_i = -(\delta z_i)/(\sin 2\theta_i) - \cot 2\theta_i (\delta z_i/\sin 2\theta_i)^2 + 0(\delta z_i^3)$.
Also, from (2.3.1),

$$E(\delta z_i) = 0,$$

and to order δt,

$$E(\delta z_i^2) = z_i(1 - z_i)\delta t/2N_e = (\sin 2\theta_i)^2 \delta t/8N_e$$

and

$$E(\delta z_i \delta z_j) = -(\cos \theta_i \cos \theta_j)^2 \delta t/2N_e.$$

Thus, to order $1/N_e$,

$$M_i(\underline{\theta}) = -\cot 2\theta_i/8N_e$$
$$V_{ii}(\underline{\theta}) = 1/8N_e$$
$$V_{ij}(\underline{\theta}) = -\cos^2\theta_i \cos^2\theta_j/(2N_e \sin 2\theta_i \sin 2\theta_j) \quad \text{for } i \neq j$$
$$= -\cot \theta_i \cot \theta_j/8N_e$$

and $(\sum\limits_{i=1}^{k} \delta(\cos^2\theta_i)) = 0.$

These may be substituted into (2. 3. 4) to give the required diffusion equation. Although the variance matrix is precisely that required for Brownian motion on a sphere (see Edwards (1971)) we have also the drift terms $M_i(\underline{\theta})$. The mean drift increases exponentially in time and is directed towards the edges of the space, but is of order t/N_e while the standard deviation is of order $(t/N_e)^{\frac{1}{2}}$. The drift term causes effects near the edges of the space, where the absorption rate of alleles is greater than that given by the Brownian motion alone.

In order for the drift to be negligible, we require

$$\left|(t/8N_e)\cot 2\theta_i\right| << (t/8N_e)^{\frac{1}{2}}$$

or

$$\tan 2\theta_i >> (t/8N_e)^{\frac{1}{2}}. \tag{2.3.5}$$

If $t/N_e \le 0.1$ this reduces to $\theta_i >> 3^0$ or $z_i >> 0.3\%$ for all alleles. The drift term may then be ignored and we have Brownian motion on a $(1/2^k)$th part of a unit sphere. Thus provided the number of generations elapsed is less than one tenth of the variance effective population size, which is not too stringent a condition whether we consider American Indian villages over the last 500 years or larger populations over the last 50,000, the mean drift may be ignored except at the extreme edges of the space. All that is necessary is that the loci used are truly polymorphic, and that we do not have absorption of some alleles during the process of evolution.

Note that, if $\delta\theta$ is the angular distance travelled in time δt,

$$E(\delta\theta^2) = E((\cos^{-1}[\sum_{i=1}^{k}(z_i(z_i + \delta z_i))^{\frac{1}{2}}])^2) \text{ from (2.3.3)}$$

$$= E((\cos^{-1}[1-(1/8)\sum_{i=1}^{k}(\delta z_i^2/z_i)])^2) = E(\sum_{i=1}^{k}(\delta z_i^2/z_i)/4) \text{ [to order } \delta z_i^2]$$

$$= (k - 1)\delta t/8N_e, \tag{2.3.6}$$

and thus we have a mean square distance proportional to $(k - 1)$, the number of independent dimensions.

Edwards (1971) makes a further transformation; the stereographic projection of the $(1/2^k)$-sphere into a (k-1)-dimensional Euclidean space. The diffusion is (k-1)-dimensional, but the chord, or angular, distances can be embedded only in a Euclidean space of k dimensions. Thus these pairwise distances could not have arisen under a model of Brownian motion in a Euclidean space. In practice the spread of populations on the sphere in the k-th dimension is usually small, but the stereographic projection provides an explicit space of the required dimension. Under the action of r. g. d. the populations approximately perform Brownian motion in this space, and it is in this projected space that population distances should be measured.

If

$$y_i = \frac{2(z_i^{\frac{1}{2}} + k^{-\frac{1}{2}})}{(1 + \sum\limits_{i=1}^{k} (z_i/k)^{\frac{1}{2}})} - k^{-\frac{1}{2}} \quad (i = 1, \ldots, k) \qquad (2.3.7)$$

where z_i $(i = 1, \ldots, k)$ are the gene frequencies at a k-allele locus, then the point $\underline{y}$ performs approximate Brownian motion in the (k-1)-dimensional space $\sum\limits_{i=1}^{k} y_i = k^{\frac{1}{2}}$. This is due to the orthomorphic nature of the stereographic projection, which results in the property, stated by Edwards (1971) and proved by Thompson (1972), of spherical contours for the likelihood function for sufficiently large multinomial samples. Random genetic drift is then repeated multinomial sampling.

The stereographic projection results in further distortion near the edges of the space. Edwards (1971) gives the upper bound to the 'scale factor' by which distances may be increased; namely $2/(1 + k^{-\frac{1}{2}})$ (or 1.17, 1.27, 1.33 for k = 2, 3, 4 respectively), but this occurs only at the extreme vertices of the space and does not arise in practice. More generally we may take an orthogonal transformation of $\underline{y}$ to obtain (k - 1) orthonormal coordinates x_j (j = 1, $\ldots$, (k-1)) in the space $\sum\limits_{i=1}^{k} y_i = k^{\frac{1}{2}}$. As for $\delta\theta_i$ above we may then consider the means and variance of the transformed diffusion. We again have a drift term of order (t/N_e) which is negligible under the same condition (2.3.5). Under this same condition, although the probability of absorption of alleles during the course of evolution may be non-negligible, the distortion of the remainder of the distribution due to absorption effects is small.

Further $cov(\delta x_i, \delta x_j) = 0$ to order $\delta t/N_e$; the variance of the diffusion does however depend on the distance from the centre of the projected space (the point $z_i = \cos^2\theta_i = 1/k$ for each i). An expression for $E(\delta x_i^2)$ may be rigorously derived, but the required result is more readily obtained as follows.

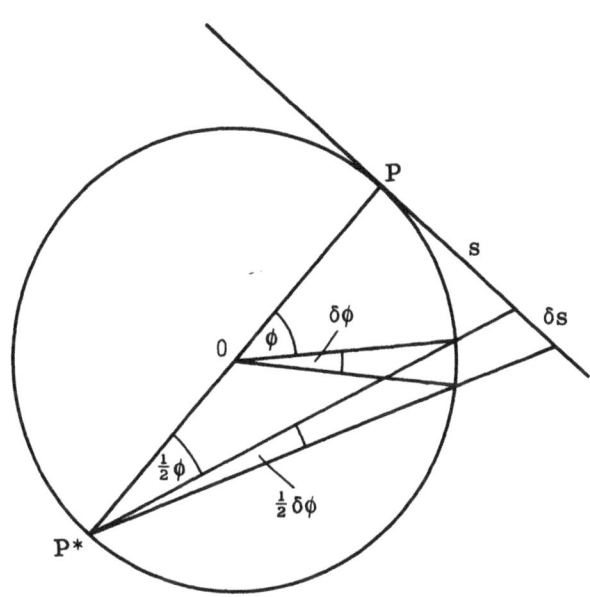

Fig. 2.3(a). The variance of the diffusion after stereographic projection from the point P*.

From Fig. 2.3(a), $s = 2\tan(\phi/2)$ and the distance δs in the projected space corresponding to small angular distance $\delta\phi$ is $\delta s = \sec^2(\phi/2)\delta\phi$. Hence, from (2.3.6), $E(|\delta\underline{x}|^2) = \sec^4(\phi/2)(k-1)\delta t/8N_e$. Thus, by the symmetry of the diffusion and the orthomorphic nature of the stereographic projection, we locally have a Brownian motion with variance

$$\sigma^2(\underline{x}) = \sec^4(\phi(\underline{x})/2)/8N_e \quad \text{per generation} \qquad (2.3.8)$$

27

in the neighbourhood of a point $\underline{x}$ at an angular distance $\phi(\underline{x})$ from the centre of the space. [In k dimensions $\sec^2(\phi(\underline{x})/2) \leq 2/(1 + k^{-\frac{1}{2}})$.] Thus if at time 0 the position of a population is known to be $\underline{x}(0)$ then after a time t, small relative to N_e, each component $x_i(t)$ of $\underline{x}(t)$ is Normally distributed;

$$N(x_i(0), \; \sec^4(\phi(\underline{x}(0))/2)t/8N_e), \quad i = 1, \ldots, (k\text{-}1),$$

and the components $x_i(t)$ are independent.

Regarding inferences from the model the inflating factor $\sec^4(\phi/2)$ has little effect, since in practice the populations are all located in some small region of the space. The only effect is to inflate all squared distances, at this given locus, by this same amount. If required the factor may be corrected for, for any given set of data, by for each given locus scaling the position vectors $\underline{x}$ of all populations by the factor $\sec^2(\phi/2)$ relevant to that region of the projected space corresponding to the observed allele frequencies; equivalently the pairwise population distances may be scaled (4.6). There is however little to be gained by making this correction; the factor is usually small, k often being only 2 or 3 for all loci. The relevant value of $\sec^4(\phi/2)$ does not normally differ significantly between loci, and the only effect is to inflate time estimates by this amount; or, since we find that times are measured only in units of N_e (3.1), N_e is similarly deflated. Estimates of tree form and relative times are unaffected; we usually infer only relative times, and even when estimates of N_e are used to infer absolute times, the above factor will be negligible compared with other sources of error - in particular, sampling errors and uncertainty concerning N_e.

For each k_i-allele locus we have $(k_i - 1)$ independent x_j, each performing a Brownian motion as the gene frequencies change due to r.g.d. We may thus combine the $\sum\limits_{i=1}^{s} (k_i - 1)$ coordinates provided by s unlinked loci, and obtain a p-dimensional Brownian motion where

$$p = \sum_{i=1}^{s} (k_i - 1). \tag{2.3.9}$$

These p coordinates will in future be referred to as the 'projected co-ordinates' of the population. k_i should be the number of alleles present in all populations: if any alleles are absent the Brownian motion for that population proceeds in a subspace of the p-dimensional space, and failure to observe this will result in underestimation of the distance/dimension and hence of evolutionary times.

We consider finally the case of two alleles as confirmation that the approximations are adequate in practice. For $k = 2$ Kimura (1955) has given a series solution to the diffusion equation (2.1.5) in terms of Gegenbauer polynomials. This density function for z, where $(z_1(t), z_2(t)) = (z, 1-z)$, may be computed for different values of q (the initial frequency $z_1(0)$) and of $u = t/N_e$, and may be transformed to obtain the true distribution on the circle quadrant and in the projected space (Fig. 2.3(b)). In the original space we have the approximation

$$z \text{ is } N(q, q(1-q)u/2) \quad \text{and} \quad 0 \le z \le 1.$$

On the circle quadrant the approximation is

$$\theta \text{ is } N(\theta_0, u/8) \text{ where } 0 \le \theta \le \pi/2, \text{ and } \theta_0 = \cos^{-1}(q^{\frac{1}{2}}).$$

In the projected space consider $h(z) = 2^{\frac{1}{2}}(z^{\frac{1}{2}} - (1-z)^{\frac{1}{2}})/(1+2^{-\frac{1}{2}}(z^{\frac{1}{2}}+(1-z)^{\frac{1}{2}}))$. $-0.829 \le h(z) \le 0.829$ and the Brownian motion approximation is

$$h(z) \text{ is } N(h(q), u/8).$$

[In the notation of (2.3.7), $h(z) = 2^{-\frac{1}{2}}(x_1 - x_2)$.]

For $q \approx 0.5$ and $u \le 0.1$ the approximations are virtually perfect. For $q = 0.1$ they are still good (Fig. 2.3(b)). Similar diagrams for $u = 0.25$ show that we still have a good approximation for $q = 0.5$, but that at $q = 0.1$ the situation is deteriorating. Besides stabilising the variance the angular transformation also improves Normality. The accuracy of the approximations after stereographic projection is similar to that before.

These results confirm that Brownian motion with variance $1/8N_e$ per generation is an adequate model for the gene frequency variation caused by random genetic drift, and, at least for $t/N_e \le 0.1$, seems much

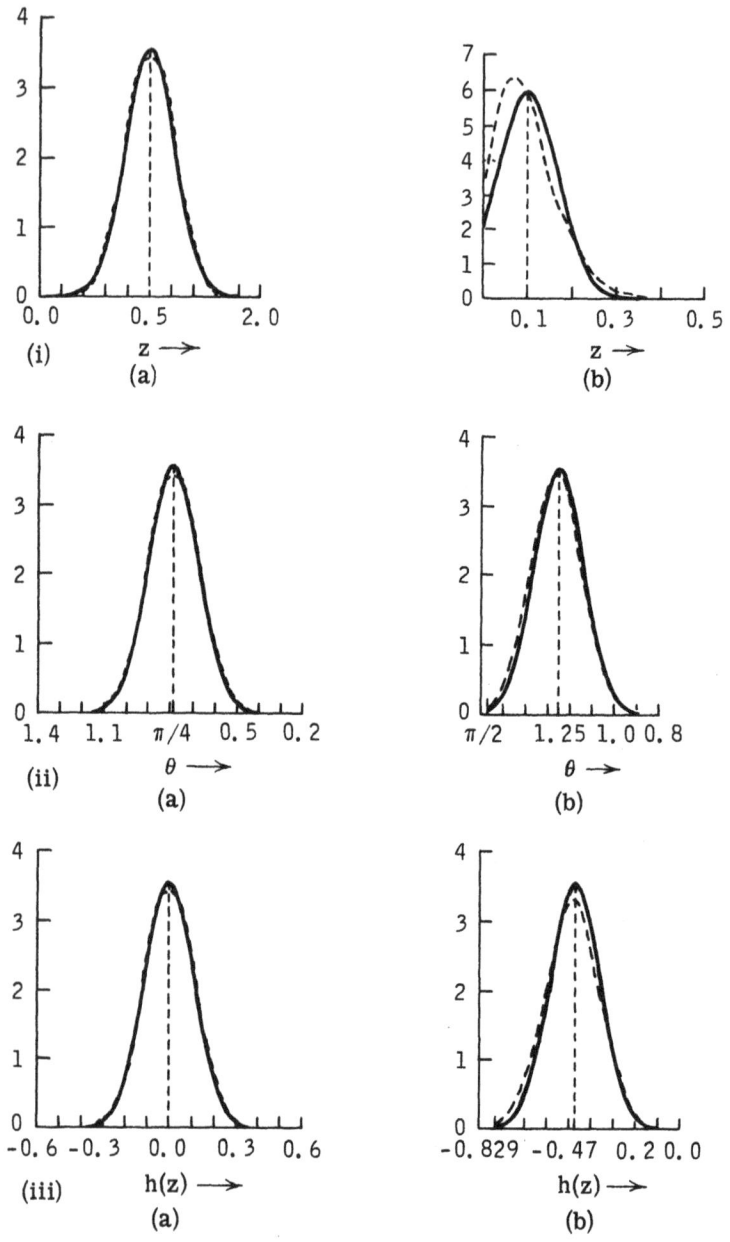

Fig. 2.3(b). The Normal approximation to the distribution for a two-allele process of random genetic drift. The broken line denotes the true density function and the continuous line the Normal approximation in each case. (i) The original gene frequency space; (ii) the representation on the surface of a sphere; (iii) the stereographically projected space. In each case we have (a) q=0.5, u=0.1 and (b) q=0.1, u=0.1. For further details see text.

better than previously expected. If all the populations are, and have remained throughout their evolution, in the same region of the projected space it should be possible to infer the evolutionary tree correctly. The final test of the model is however in the consistency and reliability of results based upon it.

2.4 THE STATISTICAL ADEQUACY OF THE MODEL

We have so far only considered the model as an approximation to r. g. d. In order to justify its application to given populations and gene loci, where differentiation may not depend wholly on r. g. d. , we must confirm that it conforms to the available data. There may be many genetic models that will fit the data, but r. g. d. is a process which has necessarily been taking place throughout history, and if the data can be fitted by a drift model alone, with plausible population and time estimates, then the adoption of the model is justified. The model must be tested independently of any likelihood inferences; these are conditional on the model (1.3). Owing to its complexity, and the number of parameters, it is not possible to test the complete model of a bifurcating tree generated by a Brownian-Yule process, but there are several aspects which may be considered individually. These are discussed with reference to the large body of data on the American Indians compiled by Post et al. (1968), with a view to examining the possibilities rather than testing the validity in this particular case. We consider inter- rather than intra-tribal variation. It has been suggested (Neel and Ward (1970)) that the evolution of the American Indian tribes may be well approximated by a tree pattern; for villages within tribes this is less likely to be so.

We must consider the aim of a statistical test of the model. A significant result means that the probability of a priori specified extreme events arising under the model is small, and that these results are then observed to have occurred; but a significance test alone should not be the cause of rejecting a model any more than of accepting it. We take the view that hypotheses may be judged only by a likelihood comparison with some alternative; a significant result may however suggest an alternative hypothesis which would provide a better explanation of the data than the proposed model (1.3). We do not consider the power of the tests below against specififed alternatives; when an alternative is contemplated a likelihood judgement and not a significance test is required. We simply consider whether the given data conforms to the proposed model; if not, we may consider the way in which it does not, and hence the ways in which the model may be inadequate.

First we consider testing agreement amongst loci as to pairwise population distances. Let $d_{ij}^{(r)}$ be the distance between populations i and j at locus r, t_{ij} the evolutionary time of this pair of populations (the time since the existence of a common ancestor), and σ^2 the variance of the Brownian motion in the projected space.

Then $d_{ij}^{(r)2}$ has distribution $2\sigma^2 t_{ij} x_{k_r - 1}^2$ and $[f_1 d_{ij}^{(r_2)2}/f_2 d_{ij}^{(r_1)2}]$ has F distribution $F(f_2, f_1)$, where $f_i = k_{r_i} - 1$. This may be tested amongst large numbers of independent population pairs: good agreement is obtained.

For any two gene loci and two disjoint populations pairs

$$P(d_{ij}^{(1)} > d_{i'j'}^{(1)} \quad \text{and} \quad d_{ij}^{(2)} < d_{i'j'}^{(2)}) = p_1(1 - p_2), \qquad (2.4.1)$$

where $p_l = P(F(k_l - 1, k_l - 1) < r)$ for $l = 1, 2$, and $r = (t/t')$ the ratio of the evolutionary times for the two pairs of populations. $p_1(1 - p_2) \to 0$ as $r \to 0$ or $r \to \infty$, as is intuitively required. But suppose $k_1 = k_2 = 2$ (values which often occur), then

$$p_1(1 - p_2) = (4/\pi^2) \tan^{-1}(r^{\frac{1}{2}}) (\pi/2 - \tan^{-1}(r^{\frac{1}{2}})).$$

This takes its maximum value of 0.25 at $r = 1$, but even for $r = 2$ the value is 0.24. Thus we can expect very little agreement in the rankings of pairwise population distances given by the different loci: certainly we do not have agreement in practice.

The inverse of the coefficient of variation of $d_{ij}^{(r)2}$ is

$$\nu = [((\text{var}(d_{ij}^{(r)2}))^{\frac{1}{2}})/E(d_{ij}^{(r)2})]^{-1} = (\tfrac{1}{2}(k_r - 1))^{\frac{1}{2}}. \qquad (2.4.2)$$

If we have a large number of independent population pairs and several loci of equal k we can, for each pair, obtain an estimate of ν. In practice a good fit to the value (2.4.2) is obtained. For example, for

32

$k = 2$ $\nu = 0.7071$, and for eight population pairs over five loci we have values with mean 0.695 and standard deviation 0.04. This test of the variance of the squared distance is a test of the 4th means of the original Normal distribution. If there are sufficient data we may test this Normal distribution completely. If $x_i^{(q)}$, $x_j^{(q)}$ are the projected coordinates of populations i and j in dimension q

$$(x_i^{(q)} - x_j^{(q)}) \text{ is distributed as } N(0, 2\sigma^2 t_{ij}) \text{ for } q=1, \ldots, p.$$

Thus for each population pair we may, if q is sufficiently large, test the fit to a Normal distribution of zero mean and unknown variance. We may then see whether some loci consistently provide the extreme values, and whether the approximation to Normality can be improved by grouping some alleles within a locus (2.2). Available samples are small, but tests again prove satisfactory apart from for a few loci (for example Duffy) which consistently provide smaller values of $|x_i^{(q)} - x_j^{(q)}|$ than do the others. Although this may be an indication of stabilising selection, this cannot be confirmed without further evidence.

That the data conform to any model does not imply that the model is correct. As a comparison the above considerations were applied also to intra-tribal differentiation. Even in this case often no substantial disagreement with the model was found - probably since sampling was often the dominant factor, and sample sizes often similar (see below). However in several cases the agreement was significantly less good than for the inter-tribal variation. While no definite conclusions should be drawn from this, it is an indication that the above tests may be sufficient to enable us to distinguish different causes of genetic variation, even with currently available data.

The final case where statistical tests may be of use is in investigating the effect of sampling. The theoretical aspects of sampling are considered in section 5.3. The gene frequencies used to reconstruct evolutionary trees are usually estimated from population samples which are unfortunately often small. Unless sample sizes for a given population differ widely between loci, in which case the observed squared distances may vary inversely, none of the above tests will distinguish between variation due to sampling and to r.g.d. since both provide distributions which are approximately Normal in the projected space (5. 3). On the assumption of sample sizes $m_1^{(r)}$ and $m_2^{(r)}$ from two populations, the distribution of the squared distance between them due to sampling alone is $[[(1/m_1^{(r)}) + (1/m_2^{(r)})]/8]\chi^2_{k_r-1}$ at locus r (see 5. 3). For different loci and disjoint population pairs we may plot observed distances against those expected under sampling alone. On the same diagram we may plot the two-unit χ^2-support limits (given by Edwards (1972: pp. 187, 227)), as shown in Fig. 2. 4. (The scale factor in the projected space has little

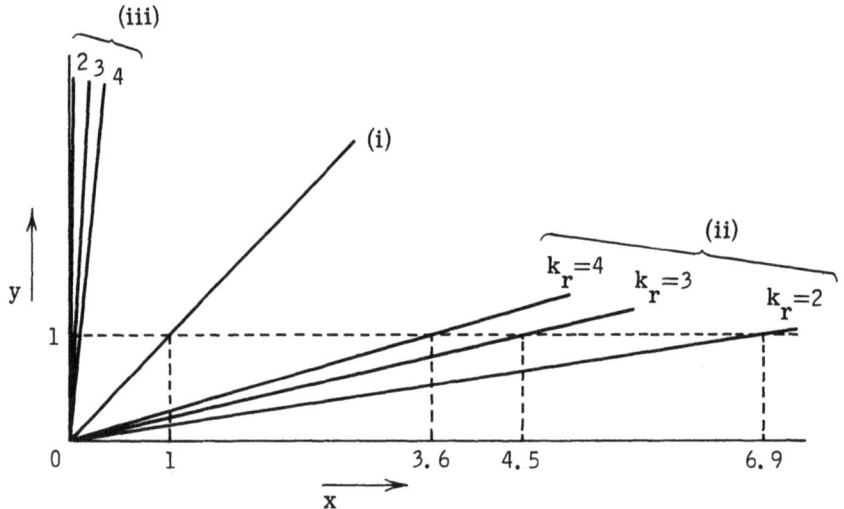

Fig. 2. 4. The comparison of sampling and observed distances.
x = observed distance between populations i and j
at locus r. y = distance expected under sampling
alone = $(k_r - 1)[(1/8m_i^{(r)}) + (1/8m_j^{(r)})]$. x/y is
χ^2_f/f, where $f = (k_r - 1)$. (i) The line $x = y$;
(ii) upper two-unit support limits for $k_r = 4, 3, 2$
respectively; (iii) lower two-unit support limits for
$k_r = 2, 3, 4$ respectively.

effect on the expected sampling distance.) For the European data of section 5. 1 we find that between several population pairs there is little evidence for significant r. g. d. , apart from for the ABO, P and PGM loci. In many cases distances can be explained by sampling alone, but this is simply because the effect of drift is too small to be reliably detected except at those loci for which large samples are available. For most other loci the correct proportion of distances (about 5%) are significantly too small, and 12% significantly too large. There seems to be significant variation amongst loci, but sample sizes are often not stated; the Duffy and Lutheran systems give significantly smaller distances.

Thus we conclude that no statistical test has provided evidence for the rejection of a Brownian motion model, and, as the simplest available adequate model for a genetic process which is known to be taking place, we are justified in adopting it. If population distances can be explained by sampling alone, then, although r. g. d. is necessarily taking place, we cannot accurately measure its effect. The reconstructed trees based on such data, and unfortunately on most currently available data, may not be accurate. However, the statistical validity of a Brownian motion model as a description of the observed data may be checked by the above tests of Normality. When large-sample data are available the true population positions are accurately known, and compatibility with Normality confirms the validity of an assumption of differentiation due to random genetic drift. The parameters of evolution may then be reliably estimated. With large-scale sampling of the more recently discovered blood groups more suitable data are rapidly becoming available.

3·The likelihood approach

3.1 THE MULTIVARIATE NORMAL MODEL

We have now certain functions of the population gene frequencies which perform p independent Brownian motions as the populations evolve in time under random genetic drift. The evolutionary tree is assumed to have been formed by a series of bifurcating splits, and it is further assumed that at some point in the past there existed a single common ancestor of the group of populations under consideration. The data are the observed functions of the gene frequencies in the n genetically distinct populations; say

$$\underline{\underline{x}} = \{x_i^{(q)}, \ i = 1, \ldots, n, \ q = 1, \ldots, p\}, \ n \geq 2, \ p \geq 1.$$

It is assumed that the population coordinates are known; that is, that sampling errors are negligible. The aim is to infer the 'labelled history', F, of the populations (2.1).

Given the form of tree, F, let s_r be the time ago of the (n-r)th split; $0 \leq s_1 \leq \ldots \leq s_{n-1}$, $\underline{s} = (s_1, \ldots, s_{n-1})$. Then F and $\underline{s}$ specify the complete history. The parameters of the model are F, $\underline{s}$, the position $\underline{x}_0$ of the initial ancestor $(\underline{x}_0 = (x_0^{(q)}, q = 1, \ldots, p))$, and the variance per generation of the Brownian motion σ^2, assumed to be the same for all populations and in all dimensions (2.2). If $\underline{x}^{(q)} = (x_i^{(q)}, i = 1, \ldots, n)$, then given $\underline{s}$ the vectors $\underline{x}^{(q)}$ (q = 1, \ldots, p) are independent, and each is multivariate Normal, being the sum of independent diffusions along relevant arcs of the tree (Fig. 3.1(a)). Hence

$$\text{cov}(x_i^{(q)}, x_j^{(q)}) = T_{ij} = \sigma^2(s_{n-1} - s_{l(i,j)}), \ q = 1, \ldots, p \quad (3.1.1)$$

where $s_{l(i,j)}$ is the time ago of the splitting of the last common ancestor of populations i and j (Adke and Moyal (1963), Felsenstein (1968)). For example, in Fig. 3.1,

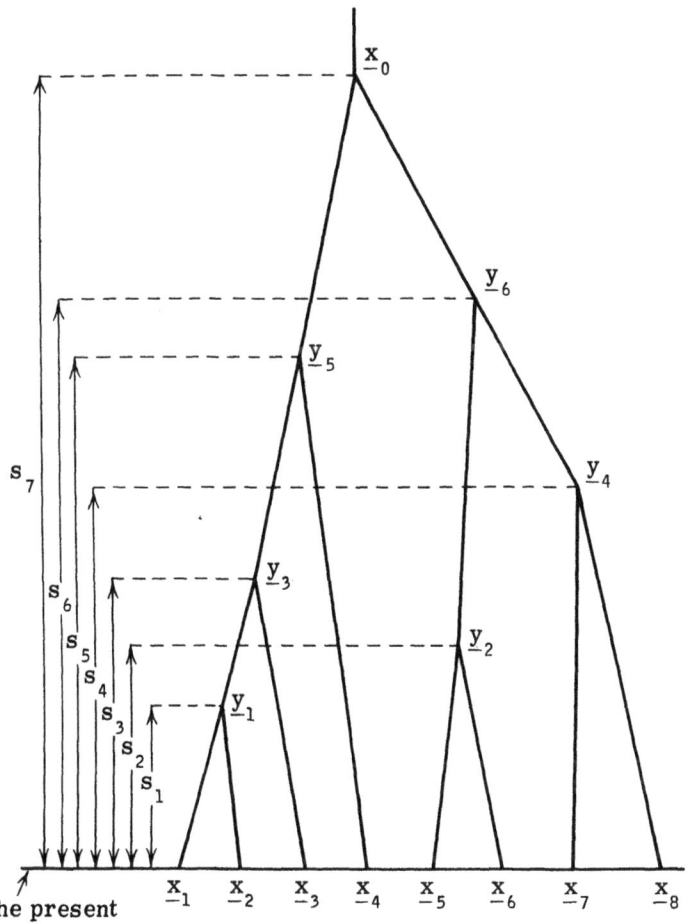

Fig. 3.1. Evolutionary tree for a group of eight populations. n = 8;
there are seven time variables and six splitting points apart from
the initial root. The topology of this tree is $(((2 + 1) + 1) + (2 + 2))$
in the notation of Harding (1971). The eight populations, labelled
1 to 8, have positions $\underline{x}_1, \ldots, \underline{x}_8$ in the projected space, and
the positions of ancestral populations at their splitting points are
$\underline{y}_1, \ldots, \underline{y}_6$. The position of the last common ancestor for this
group of populations, or the root of the tree, is $\underline{x}_0$. The time of
splitting of this ancestor, or total evolutionary time of the populations, is s_7. The form F of the tree is specified by the topology,
the labelling of final nodes, and the ordering $0 \leq s_1 \leq s_2 \leq \ldots \leq s_7$

$l\,(2,\ 3) = 3$ since s_3 is the relevant splitting time for populations $\underline{x}_2$ and $\underline{x}_3$.

Similarly, $l\,(5,\ 7) = 6$, $l\,(2,\ 7) = l\,(3,\ 6) = 7$ etc. Also $l\,(i, i) = 0$ for each i, and $s_0 = 0$. Also $E(\underline{x}_i) = \underline{x}_0$, where $\underline{x}_i = (x_i^{(q)}, q = 1, \ldots, p)$, and $\underline{x}_i \neq \underline{x}_j$ for $i \neq j$. Thus

$$f(\underline{\underline{x}}\,|\underline{x}_0,\ \underline{s},\ \sigma^2,\ F) =$$

$$(2\pi)^{-\frac{1}{2}np}\,|T|^{-\frac{1}{2}p}\exp[-\tfrac{1}{2}\sum_{q=1}^{p}(\underline{x}^{(q)}-x_0^{(q)}\underline{1})'T^{-1}(\underline{x}^{(q)}-x_0^{(q)}\underline{1})],$$

$$(3.1.2)$$

where $\underline{1}$ is a column vector of ones, and T the covariance matrix given by F and $(3.1.1)$ (Gomberg (1966), Felsenstein (1968)). [' will always denote the transpose of a vector or matrix. $\underline{x}^{(q)}$, $(q=1, \ldots, p)$, are n-dimensional column vectors. $\underline{x}_i$, $(i = 1, \ldots, n)$, and $\underline{x}_0$ may thus more intuitively be considered as p-dimensional row vectors, although they may equally be taken also as column vectors. We shall later use the standard notation $\underline{x}.\underline{y}$ for the scalar product $\underline{x}'\underline{y}$ $[\underline{x}\underline{y}']$ of two column [row] vectors.] $(3.1.2)$ is the likelihood for the parameters.

Now $f(\underline{\underline{x}}\,|\underline{x}_0,\ \underline{s},\ \sigma^2,\ F) = f(\underline{\underline{x}}\,|\underline{x}_0,\ k\underline{s},\ \sigma^2/k,\ F)$ for all $k > 0$. Inferences may be made about $\sigma^2\underline{s}$ but not σ^2 and $\underline{s}$ separately. σ^2 is simply a scale factor; it may be taken equal to 1, in which case times are measured in units of $1/\sigma^2$ generations. The order of magnitude of σ^2 will normally be known (2.3). Then the support function $S(\underline{x}_0,\ \sigma^2\underline{s},\ F)$, (1.3), is given by

$$-2S(\underline{x}_0,\ \sigma^2\underline{s},\ F) = -2\log_e L(\underline{x}_0,\ \sigma^2\underline{s},\ F) = -2\log_e f(\underline{\underline{x}}\,|\underline{x}_0,\ \sigma^2\underline{s},\ F)$$

$$= p\log|T| + \sum_{q=1}^{p}(\underline{x}^{(q)}-x_0^{(q)}\underline{1})'T^{-1}(\underline{x}^{(q)}-x_0^{(q)}\underline{1}) + \text{constant}, \quad (3.1.3)$$

where the constant is independent of both data and parameters.

This function is to be fully investigated in Chapter 4, and so few further comments need be made at this point. However even the evaluation of $(3.1.3)$ presents difficulties. The matrix T depends on F, and, although the components of T are linear in $\underline{s}$, computation of $|T|$ and

38

T^{-1} for arbitrary F is not trivial. Felsenstein (1968, 1973) gives a method for the rapid evaluation of S. An alternative method of evaluation and, more important, a method for finding the point $(\underline{x}_0,\ \sigma^2\underline{s},\ F)$ maximising S are given in Chapter 4. T may be regarded as a general covariance matrix with linear restrictions (equality and inequality) on its components. Thus the space of T values in which (3.1.3) is to be maximised is a closed convex cone in the space of all positive semi-definite symmetric matrices. However since $E(x_i^{(q)})$ varies with q but not i we do not have the usual multivariate Normal support function: we do not have simple sufficient statistics and maximum likelihood (ML) estimates cannot be explicitly found. S has no positive infinities, and T is non-singular if and only if $s_1 > 0$ (4.6).

It is often only F that is of primary interest, although, given F, we may also wish to infer $\sigma^2\underline{s}$. The MRL for F is

$$L^*(F) = \max_{\underline{x}_0,\ \sigma^2\underline{s}} L(\underline{x}_0,\ \sigma^2\underline{s},\ F) \qquad \text{(see section 1.3)},$$

and we may compare the values of $L^*(F)$ given by different values of F. Since we maximise over the same number of parameters in each case the degrees of freedom problem does not arise and a simple two-unit support difference criterion is appropriate. In practice there are too many different tree forms for all comparisons to be made (2.1), and this is the major remaining difficulty in solution (4.5). For given F we may make inferences about $\underline{x}_0$ and $\sigma^2\underline{s}$. For each $\underline{s}$ the support surface is quadratic in $\underline{x}_0$. The MRL for $\sigma^2\underline{s}$ and F

$$L^*(\sigma^2\underline{s},\ F) = \max_{\underline{x}_0} L(\underline{x}_0,\ \sigma^2\underline{s},\ F)$$

will be considered, for given F, in the following chapter.

The internal branching points $\underline{y}$ (or $\{y_i^{(q)},\ i = 1,\ \ldots,\ (n-2)$, $q = 1,\ \ldots,\ p\}$) are random variables generated by the probability model and having distributions dependent on the parameters. These variables are not relevant to likelihood inference, but we may wish to express beliefs concerning them. Having made likelihood inferences regarding the parameters, this may be done by giving the distribution of $\underline{y}$ conditional on the data and the MI estimates of the parameters, or, since this is a multivariate Normal distribution, by giving its mean and variance. This is not an entirely satisfactory procedure since, although this is the most likely distribution for $\underline{y}$, there may be other parameter values which are only marginally less likely giving very different distributions to $\underline{y}$. However it is a reasonable procedure, provided the support surface is unimodal and well-peaked at the maximum, and provided the conditional probability density of $\underline{y}$ is a continuous function of the parameters. In this particular case the procedure is acceptable within any one tree form, but there may be alternative F having almost equal support, but giving very different values to $\underline{x}_0$ and $\sigma^2 s$ and hence to the distribution of $\underline{y}$. Thus care must be taken in the use of this approach, but in general no serious logical problems arise in applying likelihood inference to the multivariate Normal model.

3.2 THE BROWNIAN-YULE MODEL

In 3.1 we have no probability model for the formation of populations: even the restriction to a bifurcating tree may degenerate if some s_i are equal. There are $(n + p)$ parameters, and this number depends on the number of populations under consideration. Edwards (1970) introduces a Yule process, rate λ, for the formation of populations. It is assumed that we are considering all the descendants of some common ancestor. The parameters of the process are now $\underline{x}_0$, σ^2, λ, and t, where $t = s_{n-1}$ is the total evolutionary time. F and $\underline{s}^* = (s_1, \ldots, s_{n-2})$ are now random variables having probability distributions dependent on the parameters, and problems arise as to how inferences should be made. The data are $(n, \underline{x}_1, \ldots, \underline{x}_n)$; n is a random variable which conveys information about λt. Thus we cannot condition upon n, and still less upon F which specifies n; F cannot be treated as a parameter independent of the Yule process.

The discrete probability distributions $P(F|n, \lambda, t)$ and $P(n|\lambda, t)$ and the densities $f(\underline{s}^*, n, F|\lambda, t)$ and $f(\underline{x}|\underline{x}_0, \underline{s}^*, t, \sigma^2, F)$ have been given by Edwards (1970), the last being (3.1.2). $P(F|n, \lambda, t)$ is independent of λ and t, and is uniform over all F having that n ($F \in H_n$ say) (Harding (1971)). For a Yule process $f(\underline{s}^*|F, n, \lambda, t)$ is independent of F (Edwards (1970)). Thus

$$L(\underline{x}_0, \sigma^2, \lambda, t) = f(\underline{x}, n | \underline{x}_0, \sigma^2, \lambda, t)$$

$$= f(\underline{x} | n, \underline{x}_0, \sigma^2, \lambda, t)P(n|\lambda, t)$$

$$= P(n|\lambda, t) \sum_{F \epsilon H_n} f(\underline{x} | n, \underline{x}_0, \sigma^2, \lambda, t, F) \; P(F|n, \lambda, t)$$

$$\propto P(n|\lambda, t) \sum_{F \epsilon H_n} [\int \cdots \int_{0 \leq s_1 \leq \ldots \leq s_{n-2} \leq t}$$

$$f(\underline{x} | \underline{x}_0, \underline{s}^*, t, \sigma^2, F, n, \lambda) f(\underline{s}^* | n, \lambda, t) d\underline{s}^*]$$

$$(3.2.1)$$

(for given n).

    This is a function only of $\underline{x}_0$, $\sigma^2 t$ and λt, and so again we only have information on relative splitting and dispersion rates. Again we may take $\sigma^2 = 1$. Edwards (1970) suggests the restriction $t = 1$, but this is less natural, particularly since we may wish to consider subsets of a group of populations.

    Then (3.2.1) may be rewritten

$$L(\underline{x}_0, \sigma^2 t, \lambda t) = P(n|\lambda t) \sum_{F \epsilon H_n} \int \cdots \int_{0 \leq s_1 \leq \ldots \leq s_{n-2} \leq t} [f(\underline{x}|\underline{x}_0, \sigma^2 \underline{s}^*, \sigma^2 t, F)$$

$$f(\underline{s}^* | n, \lambda, t) d\underline{s}^*] \qquad (3.2.2)$$

and, in theory, ML estimates of $\underline{x}_0$, $\sigma^2 t$ and λt may be made. F then has probability distribution

$$P(F|\underline{x}, \underline{x}_0, n, \sigma^2 t, \lambda t) \propto f(\underline{x}|\underline{x}_0, \sigma^2 t, \lambda t, n, F)P(F|n, \lambda t)$$

$$\propto f(\underline{x}|\underline{x}_0, \sigma^2 t, \lambda t, n, F) \text{ (as a function of F).} \quad (3.2.3)$$

A maximum probability estimate of F is the F maximising this when ML estimates are substituted for the parameters.

    However, firstly this is not a feasible procedure. The sum in (3.2.2) is over $n! (n - 1)! / 2^{n-1}$ labelled histories (2.1.1). Adke and Moyal (1963) give an iterative (over n) differential equation for the characteristic function of $f(\underline{x}, n | \underline{x}_0, \sigma^2, \lambda, t)$ but no general formulae are obtainable other than as a multiple integral and sum. Secondly, since F is a discrete variable, this procedure seems inadequate. (3.2.2) is the sum of likelihood functions, one for each F, having their maxima at different points; it may certainly be multimodal.

If some F were observed to have occurred we would have

$$L(\underline{x}_0, \sigma^2 t, \lambda t) = f(\underline{x}, F, n | \underline{x}_0, \sigma^2 t, \lambda t)$$
$$= f(\underline{x} | \underline{x}_0, \sigma^2 t, \lambda t, n, F) \; P(F|n) \; P(n|\lambda t),$$

and we would maximise this function to obtain ML estimates conditional on the given data, including F. We could then give the distributions of internal node coordinates and times conditional on the data, F and these ML estimates (cf. section 3.1). Thus instead of making overall estimates, which involves averaging over F, we may instead consider $f(\underline{x}, n, F | \underline{x}_0, \sigma^2 t, \lambda t)$ as a 'predictive likelihood', or, for fixed n, consider $G_F(\underline{x}_0, \sigma^2 t, \lambda t) =$

$$\underset{0 \le s_1 \le \ldots \le s_{n-2} \le t}{\int \int \ldots \int} f(\underline{x} | \underline{x}_0, \sigma^2 \underline{s}^*, \sigma^2 t, F) f(\underline{s}^* | n, \lambda, t) d\underline{s}^* \, . \; P(n|\lambda t) \quad (3.2.4)$$

$$= \underset{0 \le u_1 \le \ldots \le u_{n-2} \le 1}{\int \int \ldots \int} f(\underline{x} | \underline{x}_0, \sigma^2 t, \underline{u}^*, F) f(\underline{u}^* | n, \lambda t) d\underline{u}^* \, . \; P(n|\lambda t)$$

where $u_i = s_i/t$, $i = 1, \ldots, (n-2)$.

Note that

$$L(\underline{x}_0, \sigma^2 t, \lambda t) = \underset{F \in H_n}{\sum} G_F(\underline{x}_0, \sigma^2 t, \lambda t). \quad (3.2.5)$$

G_F is strictly neither a probability nor a likelihood, but we choose F, $\underline{x}_0$, $\sigma^2 t$, and λt jointly to maximise G_F. This is similar to Fisher's approach to prediction problems (1.3), and seems a reasonable procedure since $\{G_F, F \in H_n\}$ must convey the information on $\underline{x}_0$, $\sigma^2 t$, λt and F contained in the data.

In practice even the detailed examination of G_F is not feasible due to the multiple integration required to evaluate it. Its properties may be examined by direct evaluation for small n. This has been done extensively for $n = 3$ and to some extent for $n = 4$ (3.3). Although this provides some idea of possible properties of G_F, and counter-examples of hoped-for properties, it does not further the use of this model in the analysis of actual data.

G_F does have one important property for every F: namely it has (at least) one internal maximum at $(x_0, \sigma^2 t, \lambda t)$ with $0 < \lambda t < \infty$, $0 < \sigma^2 t < \infty$, $-\infty < x_0^{(q)} < \infty$ $(q = 1, \ldots, p)$. For if $n > 2$ $P(n|\lambda t) \to 0$ as $\lambda t \to 0$ or $\lambda t \to \infty$ the other part of (3.2.4) remaining bounded. Provided the populations are distinct $f(\underline{x}|x_0, \sigma^2 t, \sigma^2 \underline{s}^*, F) \to 0$ as $\sigma^2 t \to 0$ or ∞ and as $x_0^{(q)} \to \pm\infty$ for any q, uniformly in $\underline{s}^*$. Hence the integral with respect to (w.r.t.) the probability density of $\underline{s}^*$ converges to zero. Further G_F is everywhere finite and non-negative, and so has an internal maximum. Thus the problems encountered in using G_F are computational rather than theoretical. Previously it has been thought that singularities arise, and this has been a reason for not considering the Yule process (Cavalli-Sforza and Edwards (1967) and Felsenstein (1973)), but when random variables and parameters are correctly distinguished this is not the case.

3.3 THE CASE OF THREE POPULATIONS

We consider here the solution for three populations, both with and without a Yule process. This provides some insight into properties of the support surface and the problems involved in the general case.

Let $\underline{x}_1$, $\underline{x}_2$ and $\underline{x}_3$ be the positions of the three populations.
Let $s_2 = t$, $s_1 = u s_2$ and $\sigma^2 = 1$ where (s_1, s_2) is the time vector defined in 3.1 (Fig. 3.3(a)).

$$\text{Let } \underline{h} = (\underline{x}_3 - \tfrac{1}{2}(\underline{x}_1 + \underline{x}_2)) = (\underline{x}_3 - \tilde{\underline{x}})$$
$$h^2 = \underline{h}.\underline{h} = \|\underline{h}\|^2$$
$$d^2 = \|\underline{x}_1 - \underline{x}_2\|^2 > 0$$
and $\quad D = (\underline{x}_3 - \underline{x}_1).(\underline{x}_3 - \underline{x}_2).$
Thus $\quad h^2 = D + d^2/4.$

$\qquad\qquad\qquad\qquad\qquad\qquad\qquad$ (3.3.1)

Further let $K = D/d^2$, the root of the tree be $\underline{x}_0$,

$$c_0 = (\underline{x}_1 - \underline{x}_0).(\underline{x}_2 - \underline{x}_0)$$
and $\quad d_{30}^2 = \|\underline{x}_3 - \underline{x}_0\|^2.$

Let F_i be the tree form $((x_j, x_k), x_i)$ in obvious notation $(i = 1, 2, 3)$. These are the only labelled histories for $n = 3$ and w. l.o.g. we consider $F = F_3$. The covariance matrix for the tree of Fig. 3.3(a) is

$$T = \begin{pmatrix} t & t(1-u) & 0 \\ t(1-u) & t & 0 \\ 0 & 0 & t \end{pmatrix} \quad \text{with} \quad |T| = t^3 u(2-u). \qquad (3.3.2)$$

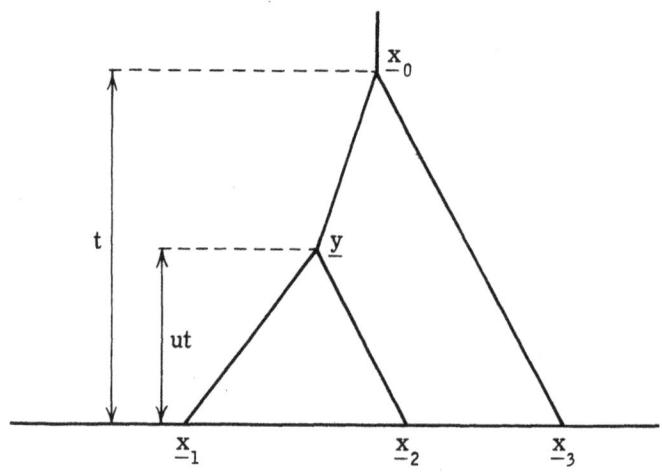

Fig. 3.3(a). Three population tree of form $F_3 = ((x_1, x_2), x_3)$.

The points x_1, x_2 and x_3 lie in a plane, and w. l. o. g.

$$x_1 = (-\tfrac{1}{2}d, 0, \ldots, 0) \quad \text{and} \quad x_2 = (\tfrac{1}{2}d, 0, \ldots, 0).$$

This fixes the location and scale and we consider the likelihood when $x_3 = (\omega_1, \omega_2, 0, \ldots, 0)$, as ω_1 and ω_2 vary.

(i)   The ME solution

One aim in obtaining an explicit likelihood solution for this simple case is to explore the often expressed hope that the ME solution provides an approximation to the ML one (1.1, 1.4). We recall therefore the

44

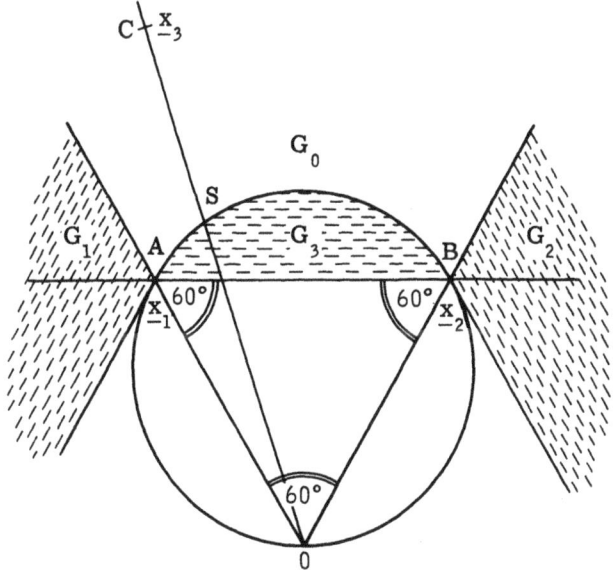

Fig. 3.3(b). Diagrammatic representation of the Steiner solution for three, necessarily coplanar, points.

Steiner solution for three points in a plane (Thompson (1973a)), shown in Fig. 3.3(b). The shortest interconnecting network is formed by joining A, B and C to the Steiner point S where if

$$x_3 \in G_1, \text{ angle } CAB > 120°, S = A$$
$$x_3 \in G_2, \text{ angle } CBA > 120°, S = B$$
$$x_3 \in G_3, \text{ angle } ACB > 120°, S = C$$

and if $x_3 \in G_0$ S is such that angles ASB, BSC and CSA are all 120°. That is, S lies at the intersection of the line OC and the circumcircle of the equilateral triangle ABO, the result being given by symmetry for x_3 below AB. (3.3.3)

ME produces an unrooted tree. Thus for three populations this is the complete solution.

(ii)   The trivariate Normal case

Consider now the likelihood (3.1.2) for the case $n = 3$. (u, t) is a 1-1 transformation of (s_1, s_2) and so the likelihood is unchanged by this reparametrisation (1. 3).

From (3. 1. 1), (3. 1. 3), (3. 3. 1) and (3. 3. 2)

$$-2S(\underline{x}_0, \ u, \ t, \ F_3) = 3p \log t + p \log(u(2 - u))$$
$$+ t^{-1}[(d^2 + 2uc_0)/(u(2 - u)) + d_{30}^2] \quad (3.3.4)$$

where the notation is as defined by (3. 3. 1).

Then maximisation w. r. t. $\underline{x}_0$ and t gives

$$\hat{x}_0(u) = [\underline{x}_1 + \underline{x}_2 + (2 - u)\underline{x}_3]/(4 - u)$$
$$\text{and} \quad \hat{t}(u) = [(d^2 + 2u\hat{c}_0)/(u(2 - u)) + \hat{d}_{30}^2]/3p, \qquad \left. \right\} \quad (3.3.5)$$

where $\hat{c}_0$ and $\hat{d}_{30}^2$ are the values of c_0 and d_{30}^2 when $\underline{x}_0 = \hat{x}_0(u)$. Further for a stationary point also w. r. t. u $(0 \le u \le 1)$,

$$Ku^3 + 4(K + 1)u^2 - (8K + 14)u + 16 = 0 \quad \text{or} \quad u = \hat{u}(K).$$

If $K > 2$, there is a unique root $0 < \hat{u}(K) < 1$, which gives a maximum of S.

If $K = 2$, $\hat{u}(K) = 1$ is the maximising stationary point.

If $K < 2$, there is no stationary point in $[0, 1]'$ and the maximum of S occurs at $\hat{u} = 1$.

Also $K > 2$ if and only if $h > 3d/2$.

There can never be more than one tree form giving a root $\hat{u}$ in $(0, 1)$, since for no population positions can two of the median lengths satisfy the required condition. Secondly, there may be no F_i with an internal maximum $(0 < \hat{u} < 1)$. The tree inferred, as $\underline{x}_3$ varies over the plane with $\underline{x}_1$ and $\underline{x}_2$ fixed, is shown in Fig. 3. 3(c). Some further points are of interest. For given form F_3;

(a) $\hat{x}_0(u) = \tilde{x} + k(u)\underline{h}$, from (3. 3. 5),   (3.3.6)

where $k(u) = (2 - u)/(4 - u)$. For $0 < u \le 1$, $1/3 \le k(u) < 1/2$, and $k(u)$ is a decreasing function of u. Thus $\underline{x}_0$ lies on the median of the triangle having the populations as its vertices at some point between the mid-point

46

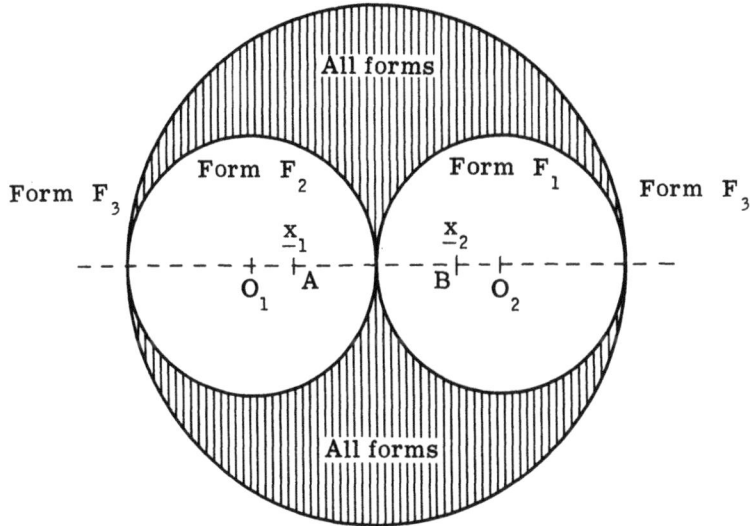

Fig. 3. 3(c).　The tri-variate Normal case.　Diagram of the tree form inferred as x_{-3} varies over the plane, with x_{-1} and x_{-2} fixed.　As x_{-3} is in each of the regions shown, the corresponding form is inferred.

$\frac{1}{2}(x_{-3} + \tilde{x})$ and the mean $\bar{x} = (x_{-1} + x_{-2} + x_{-3})/3$.

(b)　　$\hat{t}(u) = 2d^2(1 + Ku)/[3p(u(4 - u))]$ for $K \geq 2$, from (3. 3. 5),

$$(3. 3. 7)$$

and　$\log L^*(F_3) = \max_{x_{-0},\, u,\, t} [\log L(x_{-0}, u, t, F_3)]$

$$= -\tfrac{1}{2}p \log(t^3 \hat{u}(2 - \hat{u})) - 3p/2. \qquad (3. 3. 8)$$

(c)　　If $K \leq 2$, $\hat{u} = 1$, $\hat{x}_0 = \bar{x}$,

and　　$\hat{t} = (1/3p) \sum_{i=1}^{3} (x_{-i} - \bar{x}).(x_{-i} - \bar{x}) = X^2/3p,$　　　(3. 3. 9)

where X^2 is the total dispersion.

(d)　　$\hat{u}(K)$ is a decreasing function of K, and

$\hat{u} \leq \min[1, 3/(1 + K)].$　　　　　　　　(3. 3. 10)

These results together give the following

Lemma. If there is a tree with $\hat{u} < 1$ then this is the ML form. Further it gives the smallest estimate of $|T|$, the time measure of 'total dispersion', but the largest estimate of t, the total evolutionary time.

Proof. If F_i has $\hat{u} < 1$ then it is the only such F (see above). Then

$$L^*(F_i) = L(F_i, \hat{\underline{x}}_0, \hat{u}, \hat{t}) \qquad \text{where } (\hat{\underline{x}}_0, \hat{u}, \hat{t}) \text{ are estimates within } F_i$$
$$> L(F_i, \bar{\underline{x}}, 1, X^2/3p) \qquad \text{since } (\hat{\underline{x}}_0, \hat{u}, \hat{t}) \text{ are the ML estimates}$$
$$\text{given } F_i$$
$$= L(F_j, \bar{\underline{x}}, 1, X^2/3p) \qquad \text{by symmetry, for any } j$$
$$= L^*(F_j) \qquad \text{from (3.3.9).}$$

That $|T|$ is minimal for F_i follows immediately from (3.3.8) and (3.3.2). (3.3.11)

Now let d, K etc. also denote the values corresponding to form F_i. Then $\hat{t}(F_i) = 2d^2(1 + K\hat{u})/3p\hat{u}(4 - \hat{u})$ from (3.3.7), and for $j \neq i$ $\hat{t}(F_j) = X^2/3p = 2d^2(1 + K)/9p$ since $\hat{u}(F_j) = 1$, the value of $d^2(1 + K)$ being independent of the tree form for which d and K are defined (see (3.3.7), (3.3.9)). Thus $\hat{t}(F_i) < \hat{t}(F_j)$ if and only if $(1 + K\hat{u})/\hat{u}(4 - \hat{u}) < (K + 1)/3$, that is, if and only if $3/(1 + K) < \hat{u} < 1$, but, from (3.3.10) this is never the case. //

It is easily seen that (3.3.11) extends to the n population case. Writing the matrix T as tU we see from (3.1.3) that the general problem is equivalent to the minimisation of $|T| = t^n|U|$ where

$$t = (1/np)[\sum_{q=1}^{p} (\underline{x}^{(q)} - x_0^{(q)}\underline{1})'U^{-1}(\underline{x}^{(q)} - x_0^{(q)}\underline{1})],$$

but we see that this does not imply the smallest possible estimate of t.

$K > 2$ is not a stringent condition and in practice we often have a tree form with an internal root. However, the situation

$$\log L^*(F) - \log L(F, \bar{\underline{x}}, 1, X^2/3p) \geq 2$$

occurs far less often, a large value of K being required.

Finally we compare MI. and ME results. For $n = 3$ the tree forms cannot be compared since there is only one unrooted form. We can however compare the branch point positions $\underline{y}$ and S (Figs. 3.3(a), 3.3(b)).

Given $\underline{x}_1$, $\underline{x}_2$ and the parameters, $\underline{y}$ is multivariate Normal $N_p(((1 - u)(\underline{x}_1 + \underline{x}_2) + u\underline{x}_0)/(2 - u), tu(1 - u)I_p/(2 - u))$, where I_p is the $p \times p$ identity matrix.

Substituting $\hat{\underline{x}}_0(u)$ for $\underline{x}_0$ we have

$$E(\underline{y}) = \tilde{\underline{x}} + k*(u)\underline{h}, \tag{3.3.12}$$

where $k*(u) = u/(4 - u)$ [cf. (3.3.6)].

For $0 < u \le 1$, $0 < k*(u) \le 1/3$ and $k*(u)$ is an increasing function of u. $\underline{y}$ lies on the median of the triangle of the population positions at a point between $\tilde{\underline{x}}$ and $\bar{\underline{x}}$.

For $K > 2$ ($\hat{u} < 1$) the standard deviation of each $y^{(q)}$ may be quite large and to represent $\underline{y}$ by its mean position may be an over-simplification. However if we compare the forms of (3.3.12) and (3.3.3) we see that the results are qualitatively very different. Computations show that the Steiner point is rarely within two standard deviations of $E(\underline{y})$.

(iii) The Brownian-Yule model for $n = 3$

In the notation of 3.2 we have, for $n = 3$,

$$P(n = 3|\lambda t) = P_3^{(2)}(\lambda t) \text{ (the probability of 3 descendants from 2 ancestral populations)}$$

$$= 2e^{-2\lambda t}(1 - e^{-\lambda t})$$

$$f(s_1|\lambda, t) = \lambda \exp(-\lambda s_1)/(1 - \exp(-\lambda t)) \text{ for } 0 \le s_1 \le s_2 = t,$$

or $\quad f(u|\lambda t) = \lambda t \exp(-\lambda t)/(1 - \exp(-\lambda t)), \tag{3.3.13}$

for $0 \le u \le 1$, where $u = s_1/t$. Then

$$G_F(\underline{x}_0, \sigma^2 t, \lambda t) = P_3^{(2)}(\lambda t) \int_0^t f(s_1|\lambda, t) \; f(\underline{x}|\underline{x}_0, s_1, t, \sigma^2, F) \; ds_1$$

$$= \int_0^1 (\lambda t) \exp(-\lambda t(2 + u)) \; f(\underline{x}|\underline{x}_0, \sigma^2 t, u, F) \; du$$

$$= \int_0^1 r(u, \lambda t) \; f(\underline{x}|\underline{x}_0, \sigma^2 t, u, F) \; du, \tag{3.3.14}$$

and $\quad f(\underline{x} \,|\underline{x}_0, \, \sigma^2 t, \, u, \, F) = (\sigma^2 t)^{-3p/2}(u(2-u))^{-\frac{1}{2}p}\exp(-\frac{1}{2}g(\underline{x}, \, \underline{x}_0, \, u)/\sigma^2 t),$

where, for $F = F_3$, $g(\underline{x}, \, \underline{x}_0, \, u) = (d^2 + 2uc_0)/(u(2-u)) + d_{30}^2$ (cf. (3.3.4)),

$$= d^2/2u + 2d_{\tilde{0}}^2/(2-u) + d_{30}^2, \qquad (3.3.15)$$

where $d_{\tilde{0}}^2 = \|\underline{x}_0 - \frac{1}{2}(\underline{x}_1 + \underline{x}_2)\|^2 = (\underline{x}_0 - \tilde{\underline{x}}) \cdot (\underline{x}_0 - \tilde{\underline{x}})$.

For given $\underline{x}$, G_F may be numerically investigated, but more general conclusions may also be drawn. As in 3.2 G_F has at least one internal maximum w.r.t. $\underline{x}_0$, $\sigma^2 t$ and λt for each F. G_F is a mixture of the multivariate Normal likelihoods of 3.3(ii), or more generally 3.1, w.r.t. the function $r(u, \, \lambda t)$, $(0 \leq u \leq 1)$.

$[2d_{\tilde{0}}^2/(2-u)] + d_{30}^2$ is minimised, for each u, by

$$\underline{x}_0 = \tilde{\underline{x}} + k(u)\underline{h}.$$

Taking axes in the direction of $\underline{h}$ and in $(p-1)$ orthogonal directions, it may be seen that G_F is unimodal w.r.t. components of $\underline{x}_0$ in directions orthogonal to $\underline{h}$ and that

$$\hat{\underline{x}}_0 = \tilde{\underline{x}} + k\underline{h} \quad \text{for some } k. \qquad (3.3.16)$$

Thus we have

$$\begin{aligned}
G_F^*(k, \, \sigma^2 t, \, \lambda t) &= G_F(\tilde{\underline{x}} + k\underline{h}, \, \sigma^2 t, \, \lambda t) \\
&= \int_2^1 r(u, \, \lambda t)(\sigma^2 t)^{-3p/2}(u(2-u))^{-\frac{1}{2}p} \\
&\quad \exp[-\frac{1}{2}(d^2/2u + h^2(2k^2/(2-u)+(1-k)^2))/\sigma^2 t]du \\
&= \int_0^1 r(u, \, \lambda t) \, s(u, \, k, \, \sigma^2 t)du \quad \text{with } r(u, \, \lambda t) > 0. \\
&\hspace{8cm} (3.3.17)
\end{aligned}$$

For each u $s(u, \, k, \, \sigma^2 t)$ is unimodal in k and maximal at

$$k = \hat{k}(u) = (2-u)/(4-u) \qquad (\text{cf. } (3.3.6)),$$

and

$$\frac{\delta G_F^*}{\delta k} = \int_0^1 r(u, \, \lambda t) \, \frac{\delta s(u, \, k, \, \sigma^2 t)}{\delta k} \, du.$$

Hence G_F is maximal w.r.t. k at $\hat{k}$, where

$$\min_{u \in [0, 1]} \hat{k}(u) < \hat{k} < \max_{u \in [0, 1]} \hat{k}(u)$$

or

$$1/3 < \hat{k} < 1/2. \tag{3.3.18}$$

In practice it is found that G_F is unimodal w.r.t. k for each λt, $\sigma^2 t$, and that $0.38 < \hat{k} < 0.42$ over a wide range of population positions.

Thus, for any given F, $\hat{x}_0$ lies on the median of the triangle of population positions. $\hat{x}_0$ is never $\bar{x}$ and the internal branch point y coincides with x_0 with probability zero, the probability of a zero time interval. The tree is strictly bifurcating with probability one. From (3.2.5) the overall likelihood is

$$L(x_0, \sigma^2 t, \lambda t) = \sum_{i=1}^{3} G_{F_i}(x_0, \sigma^2 t, \lambda t).$$

Even if each G_{F_i} is unimodal it is possible for L to be trimodal. Whether L is trimodal or, as is also possible, unimodal in x_0, it is not necessary for x_0 at the maximal mode to lie on any of the three medians. This may be demonstrated by a distortion of the case for three populations in an equilateral triangle.

Several results may be deduced from symmetry considerations. Again we consider the situation for fixed x_1 and x_2 with x_3 varying in a plane (Fig. 3.3(d)). As before let $x_2 = -x_1 = (\frac{1}{2}d, 0, \ldots, 0)$ and $x_3 = (w_1, w_2, 0, \ldots, 0)$. Let $K_i = \max_{x_0, \sigma^2 t, \lambda t} [G_{F_i}(x_0, \sigma^2 t, \lambda t)]$, a function of the population positions or of (d, w_1, w_2), with the form F_i as previously defined. Note that for given orientation of the populations $\sigma^2 t \propto d^2 = \| x_1 - x_2 \|^2$ and $K_i \propto d^{-3p}$.

Now $K_i(d, w_1, w_2) = K_i(d, w_1, -w_2)$ for i = 1, 2, 3, $K_1(d, w_1, w_2) = K_2(d, -w_1, w_2)$, and $K_3(d, w_1, w_2) = K_3(d, -w_1, w_2)$, for all (d, w_1, w_2). As $w_1 \to \pm \infty$ (w_2 fixed), $K_2/K_1 \to 1$. As $(w_1, w_2) \to \infty$ in any direction, $K_3 = \max. (K_1, K_2, K_3)$, and $\hat{x}_0(F_3) \to \frac{1}{2} x_3$ as in the trivariate Normal case, $(\tilde{x} = 0)$.

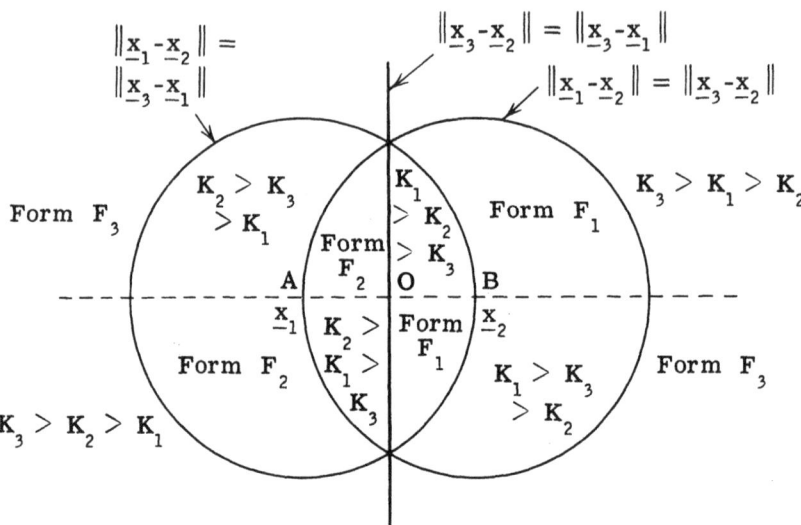

Fig. 3.3(d). The three-population Brownian-Yule case. Diagram of the tree form inferred as $\underline{x}_3$ varies over the plane with $\underline{x}_1$ and $\underline{x}_2$ fixed.

If $\|\underline{x}_1 - \underline{x}_3\| = \|\underline{x}_1 - \underline{x}_2\| = d$, $K_2 = K_3$.

If $\|\underline{x}_2 - \underline{x}_3\| = \|\underline{x}_1 - \underline{x}_2\| = d$, $K_1 = K_3$.

$$(3.3.19)$$

We find also that

$$G^*_{F_1}\ (\hat{k}(F_1),\ \sigma^2 t,\ \lambda t) \gtrless G^*_{F_2}\ (\hat{k}(F_2),\ \sigma^2 t,\ \lambda t) \text{ for each } (\sigma^2 t,\ \lambda t)$$

$$\text{as } w_1 \gtrless 0,$$

and hence $K_1(d,\ w_1,\ w_2) \gtrless K_2(d,\ w_1,\ w_2)$ as $w_1 \gtrless 0$. $(3.3.20)$

$[\hat{k}(F_i),\ \hat{\underline{x}}_0(F_i)$ denote the values $\hat{k}$ and $\hat{\underline{x}}_0$ maximising G_{F_i} for given $(\sigma^2 t,\ \lambda t).]$

From (3.3.20) the only points of equality of the K_i are the symmetry points (3.3.19) and the line $w_1 = 0$. Thus the form maximising K_i is almost surely strictly defined and consists of joining the two closest populations first. The ordering of the K_i as $\underline{x}_3$ varies is shown in Fig. 3.3(d). Note the similarities between this solution and that of Fig. 3.3(c). Computations show that in practice the form with maximum K_i has the smallest estimate of $\sigma^2 t$, in contrast to the Lemma of 3.3(ii).

In this case a quantitative comparison with ME is less easy to make: the qualitative comparison remains as before.

$$E(\underline{y}|\underline{x}, \underline{x}_0, \sigma^2 t, \lambda t) = E(E(\underline{y}|\underline{x}, \underline{x}_0, u)|\underline{x}, \underline{x}_0, \sigma^2 t, \lambda t)$$
$$= E[(2(1-u)\tilde{\underline{x}} + u\underline{x}_0)/(2-u)|\underline{x}, \underline{x}_0, \sigma^2 t, \lambda t).$$

Hence when $\underline{x}_0 = \tilde{\underline{x}} + k\underline{h}$

$$E(\underline{y}|\underline{x}, \underline{x}_0, \sigma^2 t, \lambda t) = \tilde{\underline{x}} + k\underline{h}E((u/(2-u))|\underline{x}, \underline{x}_0, \sigma^2 t, \lambda t). \quad (3.3.21)$$

Now the posterior distribution of u given $\underline{x}$ is complicated. However the prior and posterior distributions may not differ significantly, and using the distribution (3.3.13) we obtain

$$E(\underline{y}) = \tilde{\underline{x}} + (2H(\lambda t) - 1)k\underline{h}, \quad (cf. (3.3.12))$$

where

$$H(m) = \int_1^2 (e^{mu}/u)du / \int_1^2 (e^{mu})du.$$

This will approximate the true posterior mean provided the data are not in significant disagreement with a Yule process model for the times. $H(m)$ is a decreasing function of m, and for $0 \leq m < \infty$

$$\log_e 2 \geq H(m) > \tfrac{1}{2} \quad \text{and} \quad 0 < (2H(m) - 1) \leq 0.4.$$

Thus $E(\underline{y})$ will rarely be closer to $\hat{\underline{x}}_0$ than to $\tilde{\underline{x}}$ (cf. 3.3(ii)). As $\lambda t \to \infty$, $(2H(\lambda t) - 1) \to 0$ and $E(\underline{y}) \to \tilde{\underline{x}}$ as is to be expected. Using the prior distribution of u an expression for the variance of $\underline{y}$ may be similarly obtained, and hence the position of $\underline{y}$ compared with that of the Steiner point S.

Unimodality of the trivariate Normal likelihood of 3.3(ii) does not necessarily imply unimodality of G_F, even for the simple mixing function $r(u, \lambda t)$. In practice however we always find a local maximum, which is unique at least in the region of the parameter space of interest. We have already noted that the overall likelihood (3.2.5) need not be unimodal. We have also seen that G_F is unimodal w.r.t. components of $\underline{x}_0$ orthogonal to $\underline{h}$, and that for the situation with regard to k we need consider only

$1/3 < k < 1/2.$

If we consider the second derivative of (3.3.17) we find that this must be negative, the integrand being negative for every u, provided

$$h^2/\sigma^2 t < \min(2/(1 - 2k)^2, \; 3/(1 - 3k)^2).$$

For $k = 1/3$, 0.38, 0.42 and 1/2 the right hand side is 18, 35, $44\frac{1}{2}$ and 12 respectively; normally $0.38 < \hat{k} < 0.42$ (see above).

Given u, each component of $\underline{h}$ is $N(0, \sigma^2 t(2 - \frac{1}{2}u))$, and thus

$$E(h^2) < 2\sigma^2 pt, \quad E(h^2/\sigma^2 t) < 2p.$$

Thus for those $\sigma^2 t$ of interest G_F is unimodal in k, and hence in $\underline{x}_0$. Similar considerations for $\sigma^2 t$ show that again we need only consider a central range of values, and that for those $\underline{x}_0$ and λt of interest the second derivative is negative in this range. Thus we may expect unimodality w.r.t. this parameter also.

Unimodality in each parameter separately is not sufficient for a unique overall mode. To obtain such a result we require either that the equations for a stationary point have a unique root, or that any stationary point is a local maximum. However the complexity of G_F does not allow these requirements to be checked. The above considerations, together with practical results, do however lead us to the conclusion that there will usually be only one maximum in the region of interest in the parameter space.

3.4 A BIRTH AND DEATH PROCESS RESULT

A model for population splitting is assumed in order to decrease the dimensionality of the parameter space. The Yule process model may however be unrealistic. It makes no allowance for populations which have become extinct or absorbed by others. More seriously it makes no allowance for the fact that we rarely consider all the major descendants of some ancestral population. These criticisms may be met by having instead a birth and death process model, rates λ and μ respectively, for the formation and extinction of populations. Besides allowing for true extinction this model may also give a more realistic distribution for the times of split for a subset of the descendants of a common ancestor.

The standard birth and death process probabilities for the existence of n descendants (from a single ancestor) after time t are

$$p_0(t) = \mu(1 - E)/(\lambda - \mu E)$$
$$p_1(t) = (\lambda - \mu)^2 E/(\lambda - \mu E)^2$$
and $\quad p_n(t) = (\lambda/\mu)^{n-1}p_1(t)[p_0(t)]^{n-1} \quad \text{for } n \geq 2,$
where $E = \exp(-(\lambda - \mu)t)$.

$$(3.4.1)$$

These probabilities are functions of λt and μt only.

Now the main points in the derivation of the likelihood of 3.2 are that the times of split are independent of the form of tree and that, given n, each tree form in H_n is equiprobable. Suppose we define a 'significant' split as one which is finally distinguishable: that is, one that results in descendants of both daughter populations at the final time of observation.

The considerations of Harding (1971) show that a necessary and sufficient condition for $P(F|\underline{s}*, n)$ to be independent of F $(F \in H_n)$ and hence necessarily of $\underline{s}*$ is that, given n and $\underline{s}*$,

at whatever times the significant splits occurred,
each did so with equal probability in each population
then existent, regardless of the previous history.

$$(3.4.2)$$

Now in a simple birth and death process populations split independently and, at any given time, all populations have equal probability of splitting and of having final descendants. Conversely, if populations split independently of each other, only such a process can give $P(F|\underline{s}*, n)$ independent of both F and $\underline{s}*$. Thus for a birth and death process we have the required results: an explicit form for $P(\underline{s}*, n, F)$ is derived below.

Thus, as in (3.2.4), we may write

$$G_F(\underline{x}_0, \sigma^2 t, \lambda t, \mu t) = P(n|\lambda t, \mu t).$$

$$\int \int \cdots \cdots \int_{0 \leq s_1 \leq \ldots \leq s_{n-2} \leq t} f(\underline{s}*|\lambda, \mu, t, n)f(\underline{x}|\underline{x}_0, \sigma^2 \underline{s}*, \sigma^2 t, F, n)d\underline{s}*.$$

$$(3.4.3)$$

As in (3.2.5) the complete likelihood for the parameters is the sum of the

G_F over all F in H_n, but again G_F may be considered for the purpose of making inferences about F. All the parameters $(x_0, \sigma^2 t, \lambda t$ and $\mu t)$ are identifiable, but it may be that to obtain a likelihood of reasonable shape some additional constraint (for example λ/μ fixed) must be imposed. This is the case for the likelihood $p_n(t)$ for λt and μt given n descendants of a single ancestor; whereas in the unrestricted case $\hat{\mu} = 0$ (provided $n \neq 0$), if λ/μ is fixed a unimodal likelihood for λt is obtained. If such a constraint is imposed there are the same number of independent parameters as before, but the new distribution for the times may provide substantially better fit.

Thus, using the result (3.4.6) below, there is no intrinsic problem in generalising the model to a birth and death process. The likelihood theory remains the same. However, the functions involved, and hence the multiple integration of (3.4.3), are more complicated in this case and no numerical investigations of G_F have been attempted.

Theorem. In the previous notation the joint distribution

$$f(\underline{s}^*, n, F | \lambda, \mu, t) d\underline{s}^*$$

is independent of F $(F \in H_n)$ and proportional to

$$\prod_{i=1}^{n-2} p_1(s_i), \tag{3.4.4}$$

where s_i is the time ago of the $(n-i)$th significant split;

$$0 = s_0 \leq s_1 \leq s_2 \leq \ldots \leq s_{n-2} \leq s_{n-1} = t.$$

Notes. (i) The time distribution given by Edwards (1970) for the special case of a Yule process $(\mu = 0)$ satisfies the theorem.

(ii) $f(\underline{s}^* | n, F, \lambda, \mu, t) = f(\underline{s}^*, n, F | \lambda, \mu, t)/P(F | n)P(n | \lambda, \mu, t)$ and $P(F | n) = 1/|H_n|$ for all $F \in H_n$. Thus the conditional density $f(\underline{s}^* | n, F, \lambda, \mu, t)$ is independent of F and is also of the form (3.4.4).

(iii) Adke and Moyal (1963) provide a result similar to (3.4.4) for unlabelled histories descended from a single population but no direct proof is given.

Proof. First $\sum_{k=1}^{\infty} [k(p_0(s))^{k-1} p_1(s) p_k(t - s)] = p_1(t).$ (3.4.5)

This may either be proved directly for a birth and death process from (3.4.1), or we may see that it must hold for any branching process of independent increments; for to obtain one descendant at time t there must be k descendants at any given time $(t - s)$ $[0 < s < t,$ some $k \geq 1]$, and any one of these k must have one final descendant, and the remaining $(k - 1)$ lines must be extinct.

Now define $\varepsilon_{ij} = 1$ if there is a branch of the tree joining the splits at s_i and s_j and $i > j$ $(s_i \geq s_j)$, and $\varepsilon_{ij} = 0$ otherwise. Consider such a branch of the tree. The relevant population at time s_i will result in k_{ij} descendants by time s_j with probability

$$p_{k_{ij}}(s_i - s_j).$$

One of these splits in time interval δs with probability

$$\lambda k_{ij} \delta s.$$

The remaining $(k_{ij} - 1)$ must be extinct lines by the time a further interval s_j has elapsed, which occurs with probability

$$(p_0(s_j))^{(k_{ij}-1)}.$$

Further for $j = 0$ and $\varepsilon_{ij} = 1$ we must have $k_{ij} = 1$: these are the arcs joining the final splits to the present populations (Fig. 3.4).

The total probability is then the product over all independent arcs. Thus $f(s_1, \ldots, s_{n-2}, n, F | \lambda, \mu, t)$

$$= [\lambda^{n-2}/C(n)] \prod_{\substack{\varepsilon_{ij}=1 \\ 0 \leq j \leq (n-2) \\ 1 \leq i \leq (n-1)}} \sum_{k_{ij}> 0, \, k_{i0}=1} [k_{ij} p_{k_{ij}}(s_i-s_j)(p_0(s_j))^{(k_{ij}-1)}]$$

where, as in Edwards (1970), $C(n)$ is a function of n alone, and is equal to the number of distinct labellings of an unlabelled history. Thus from (3.4.1) and (3.4.5), $f(s_1, \ldots, s_{n-2}, n, F | \lambda, \mu, t)$

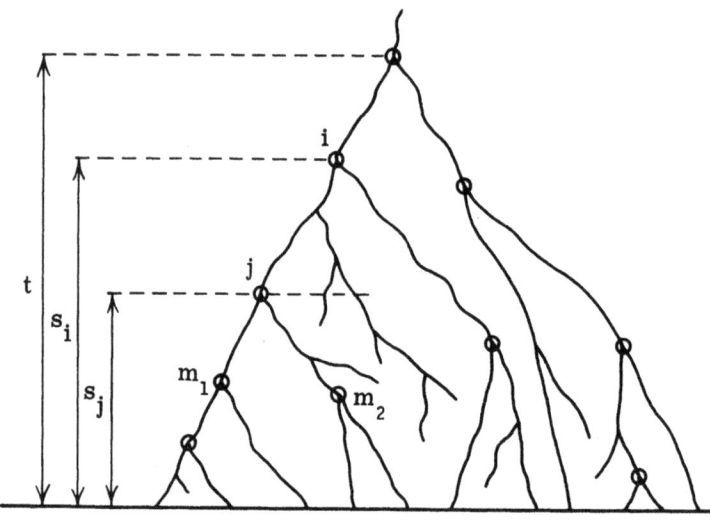

Fig. 3.4. Evolutionary tree given by a birth and death process for population formation. Significant splits are denoted by $\bigotimes$, and $k_{ij} = 3$.

$$= [\lambda^{n-2}/C(n)]. \prod_{\substack{\varepsilon_{ij} = 1 \\ 0 < j \leq (n-2) \\ 1 \leq i \leq (n-1)}} [p_1(s_i)/p_1(s_j)]. \prod_{\substack{\varepsilon_{i0} = 1 \\ 1 \leq i \leq (n-1)}} [p_1(s_i)]$$

$$= [\lambda^{n-2}/C(n)] \prod_{\substack{\varepsilon_{ij} = 1 \\ 0 \leq j \leq (n-2) \\ 1 \leq i \leq (n-1)}} [p_1(s_i)/p_1(s_j)], \text{ where } p_1(s_0) = p_1(0) = 1,$$

$$\text{and} \quad C(n) = n! \, /2^{n-1}.$$

Now for each arc (i, j) $(\varepsilon_{ij} = 1, 1 \leq j \leq (n-2))$ there are two arcs (j, m_1) and (j, m_2) (Fig. 3.4). There are also two arcs $((n-1), j_1)$ and $((n-1), j_2)$.

Thus $f(s_1, \ldots, s_{n-2}, n, F | \lambda, \mu, t)$

$$= 2^{n-1}\lambda^{n-2}[p_1(t)]^2 \prod_{j=1}^{(n-2)} [p_1(s_j)]/n! \tag{3.4.6}$$

which is the required result. //

4 · A likelihood solution

4.1 INTRODUCTION

In the previous chapter it was seen how some model for the splitting of populations, in particular the Yule process, may be included in the
probability model for the data. The purpose of this is to decrease the
dimensionality of the estimation problem, and hence, in theory, to simpli-
fy it. In spite of the attraction of a model requiring only three basic
parameters $(\underline{x}_0,\ \lambda t$ and $\sigma^2 t)$ to describe the course of human evolution
it must be accepted that the splitting model cannot be very realistic.
Further, although the dimensionality of the estimation problem is reduced
from $(n + p)$ to $(p + 2)$, and becomes independent of n, the Yule model
greatly increases the complexity of the likelihood function. It also raises
the unresolved problem of inferences concerning an unobserved discrete
random variable (1.3, 3.2). Even with the extension to the birth and
death process of section 3.4 the model for splitting times may still be
unrealistic, particularly in its assumption that all existent descendants
of some common ancestor are to be investigated.

Acceptance of a Yule process model does not introduce singularities
or other unacceptable properties into the likelihood surface. However
partly because of the non-validity of the model, partly because of the
inference problems concerning F, but mainly due to purely computational
problems, we now drop the Yule model and consider only the multivariate
Normal likelihood of section 3.1. Only when this simpler problem is fully
solved will it be possible to see how some model for population splitting
may be reasonably incorporated.

In this chapter we therefore construct and investigate a method for
the likelihood solution of the evolutionary tree problem on the basis of the
model of section 3.1. Some discussion of the theoretical approach, some
basic notation and some properties of the likelihood surface were given in

that section. As before we have contemporary gene frequency data for n populations for blood group loci providing p Brownian motion dimensions. We consider the data coordinates $\underline{\underline{x}}$ as

$$\{\underline{x}_i, \ i = 1, \ \ldots, \ n\},$$

where $\underline{x}_i = (x_i^{(q)}, \ q = 1, \ \ldots, \ p)$ or as

$$\{\underline{x}^{(q)}, \ q = 1, \ \ldots, \ p\},$$

where $\underline{x}^{(q)} = (x_i^{(q)}, \ i = 1, \ \ldots, \ n)$, a column vector (3.1).

For given basic parameters $\underline{x}_0[= (x_0^{(q)}, \ q = 1, \ \ldots, \ p)]$ and $\underline{s}$ $[= (s_1, \ \ldots, \ s_{n-1}), \ 0 \le s_1 \le \ldots \le s_{n-1}]$ and labelled history F, we have (3.1.3);

$$-2 \log L(\underline{x}_0, \ \sigma^2 \underline{s}, \ F) = p \log |T| + \sum_{q=1}^{p} (\underline{x}^{(q)} - x_0^{(q)} \underline{1})' T^{-1} (\underline{x}^{(q)} - x_0^{(q)} \underline{1}),$$

$$(4.1.1)$$

where the notation is as previously defined.

There are some preliminary points to be made. Firstly, the method will be seen to produce a maximum likelihood tree for a given labelled history, F. In theory the value of $\max\limits_{\underline{x}_0, \ \sigma^2 \underline{s}} . \ [L(\underline{x}_0, \ \sigma^2 \underline{s}, \ F)]$ should then be compared over all the $[(n! (n-1)!)/2^{n-1}]$ values of F (2.1) - a clearly impossible task. An idea on solving this problem is given in section 4.5 and used in the program that has been developed (5.1). Secondly, the gene frequency data must be contemporary. To assign data points to different times on the evolutionary tree creates singularities in the likelihood. This anomalous situation is considered in section 4.6.

The gene frequencies, which determine the $x_i^{(q)}$, are not actual population frequencies, but are estimated from population samples. However, we assume in constructing the model that $\underline{x}_i$ are known population positions. The inclusion of sampling is considered in section 5.3 but the problems are not fully solved. We note that for a Brownian motion model the likelihood must be independent of the actual coordinate system in the p-dimensional Euclidean space. Use is made of this in section 4.6, but until then the $x_i^{(q)}$ may be taken as the projected coordinates given in Chapter 2.

Finally we emphasise again the necessity of drawing a clear distinction between parameters and random variables, since this distinction is particularly important over the next few sections. Under the present model x_0, $\sigma^2 s$ and F are basic parameters and have a likelihood given the data x, being the probability density of the data given the parameters. Since there is no model for the production of populations n is a chosen constant. The internal nodes of the process have positions which are random variables, having a probability distribution given the parameters, or a conditional distribution given the data x and the parameters.

4.2 NOTATION AND PRELIMINARY FORMULAE

Before developing the method some new notation must be introduced. In place of s_k, the time ago of the $(n - k)$th split, we now consider the kth time interval ago. That is, t_k is the time between the $(n - k)$th split and the $(n - k + 1)$th. Then $t_1 = s_1$ and $t_i = (s_i - s_{i-1})$ for $i = 2, \ldots, (n-1)$, and $t_i \geq 0$. $t [= (t_1, \ldots, t_{n-1})]$ is a 1-1 transformation of s, and the likelihood is unchanged. To avoid repeated provisos we assume $t_1 > 0$ and that the populations are distinct.

In the new notation,

$$T_{ij} = \sigma^2 [\sum_{k=l(i,\,j)+1}^{(n-1)} t_k] \quad \text{(cf. (3.1.1))} \tag{4.2.1}$$

Let the number of populations existing in the kth time interval ago be n_k: then $n_k = (n - k + 1)$. Further when the data are contemporary σ^2 is non-identifiable, being simply a scale factor for times. Thus we may take $\sigma^2 = 1$, and measure time in units of $1/\sigma^2$ generations. Finally, for a given form of tree F, define,

$$H(x, x_0, t) = (x - x_0 1)'T^{-1}(x - x_0 1), \tag{4.2.2}$$

where T is the covariance matrix defined by F and t, via (4.2.1). Then rewriting (4.1.1);

$$-2S(\underline{x}_0, \underline{t}, F) = -2 \log L(\underline{x}_0, \underline{t}, F)$$

$$= p \log |T| + \sum_{q=1}^{p} H(\underline{x}^{(q)}, x_0^{(q)}, \underline{t}). \qquad (4.2.3)$$

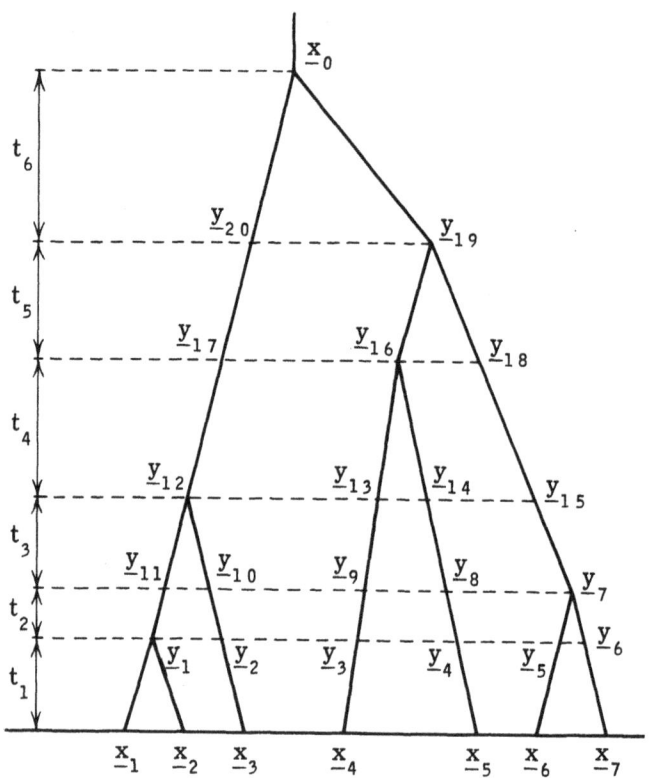

Fig. 4. 2.   Example of a tree showing type 0 (basic) and type 1 internal nodes. $n = 7$, $(n-1) = 6$, and $r_0 = \frac{1}{2}(n+1)(n-2) = 20$.

Topology of tree $= ((2 + 1) + (2 + 2))$.

Data are $\underline{x}_1, \ldots, \underline{x}_7$, where $\underline{x}_i = (x_i^{(q)}, q = 1, \ldots, p)$.

The parameters are $t_1, \ldots, t_6, \underline{x}_0$ and F.

The additional type 0 variables are $\underline{y}_1, \underline{y}_7, \underline{y}_{12}, \underline{y}_{16}$ and $\underline{y}_{19}$, and the type 1 variables are all the other $\underline{y}_i$, $1 \le i \le 20$.

[The ordering of the labels j of vectors $\underline{y}_j$ is immaterial.]

We now introduce some notation for the internal nodes of the system. There are $(n-1)$ actual internal splitting points, which we call 'type 0' (or __basic__) nodes. We consider also the 'type 1' internal nodes, which are the positions of the other populations at the instant at which some population splits. There are $\frac{1}{2}(n+1)(n-2)\ [=r_0]$ internal nodes in all, and these are denoted by

$$\underline{y} = \{\underline{y}^{(q)},\ q=1,\ \ldots,\ p\},\ \underline{y}^{(q)} = (y_i^{(q)},\ i=1,\ \ldots,\ r_0),\ \text{(Fig. 4.2)}.$$

The total number of variables is of order n^2 and if it were necessary to consider them all this approach would hardly be feasible, but we shall find that type 1 nodes are useful only as a theoretical concept and that only type 0 need be considered in computation.

Denote the conditional distribution of $\{y_i^{(q)}\}$ given the data and the parameters by

$$f(\underline{y}|\underline{x},\ \underline{x}_0,\ \underline{t},\ F) = f(y_i^{(q)};\ i=1,\ \ldots,\ r_0,\ q=1,\ \ldots,\ p\,|\underline{x},\ \underline{x}_0,\ \underline{t},\ F),$$

and the joint distribution of $\underline{x}$ and $\underline{y}$ by $f(\underline{x},\ \underline{y}|\underline{x}_0,\ \underline{t},\ F)$.

$$\text{Then } L(\underline{x}_0,\ \underline{t},\ F) = \prod_{q=1}^{p} f(\underline{x}^{(q)},\ \underline{t},\ F) = f(\underline{x}|\underline{x}_0,\ \underline{t},\ F)$$

$$= \int\!\int \ldots \int_{\underline{y}} f(\underline{x},\ \underline{y}|\underline{x}_0,\ \underline{t},\ F)d\underline{y} \qquad (4.2.4)$$

and $f(\underline{y}|\underline{x},\ \underline{x}_0,\ \underline{t},\ F) = f(\underline{x},\ \underline{y}|\underline{x}_0,\ \underline{t},\ F)/L(\underline{x}_0,\ \underline{t},\ F). \qquad (4.2.5)$

All these distributions are multivariate Normal, being independent and identical in each dimension q, and can thus be specified by the means and covariance matrices. The maximum probability estimates and the means are identical. Further, since the complete joint distribution is proportional to $\exp(-\frac{1}{2}Q)$ where Q is a quadratic form in $\underline{x}_i,\ \underline{y}_i$ $(i = 1,\ \ldots,\ n)$ and $\underline{x}_0$, the means are linear in the conditioning variables and in $\underline{x}_0$.

Let $m_i^{(q)} = E(y_i^{(q)}|\underline{x}^{(q)},\ x_0^{(q)},\ \underline{t},\ F)$ for $i = 1,\ \ldots,\ r_0$, $q = 1,\ \ldots,\ p$, and consider the arcs of the tree crossing time interval t_k. Let $\varepsilon_{ij}(k) = 1$ if $\underline{y}_i$ is the position of a population at time $\sum_{l=1}^{k} t_l$ ago and $\underline{y}_j$ is the position of the same population after time interval t_k. Otherwise let $\varepsilon_{ij}(k) = 0$. Define

$$C_k^{(q)} = \Sigma_i \Sigma_j \epsilon_{ij}(k)[y_i^{(q)} - y_j^{(q)}]^2, \text{ and } C_k = \sum_{q=1}^{p} C_k^{(q)}, \qquad (4.2.6)$$

and

$$D_k^{(q)} = \Sigma_i \Sigma_j \epsilon_{ij}(k)[m_i^{(q)} - m_j^{(q)}]^2, \text{ and } D_k = \sum_{q=1}^{p} D_k^{(q)}. \qquad (4.2.7)$$

Further let

$$M_k^{(q)} = E(C_k^{(q)} | \underline{\underline{x}}, \underline{x}_0, \underline{t}, F) \text{ and } M_k = \sum_{q=1}^{p} M_k^{(q)}. \qquad (4.2.8)$$

Note that C_k is the 'total (distance)^2 travelled in time interval t_k by all populations then existent', and that

$$M_k^{(q)} = \Sigma_i \Sigma_j \epsilon_{ij}(k) E((y_i^{(q)} - y_j^{(q)})^2 | \underline{\underline{x}}, \underline{x}_0, \underline{t}, F)$$

$$= D_k^{(q)} + \Sigma_i \Sigma_j \epsilon_{ij}(k) \text{ var}((y_i^{(q)} - y_j^{(q)}) | \underline{\underline{x}}, \underline{x}_0, \underline{t}, F) \geq D_k^{(q)}.$$

M_k is the 'mean total divergence' while D_k is the 'total divergence of the means'.

Further let $\underline{z} = \{\underline{x}, \underline{y}\}$, the total set of internal node and population positions, the vectors $\underline{z}_i = (z_i^{(q)}, q = 1, \ldots, p)$, for $i = 1, \ldots, m$ and $m = r_0 + n$, being ordered in time. Let $\underline{z}_i^*$ be the immediate ancestor of $\underline{z}_i$. Then $\underline{z}_i^*$ is either some $\underline{z}_j$ with $j < i$ or the parameter $\underline{x}_0$. Let $\{\underline{z}_{H_i}; i = 1, \ldots, (2n-1), 1 \leq H_i \leq m\}$ be the set of type 0 internal nodes $(i = 1, \ldots, n-1)$ and populations $(i = n, \ldots, 2n-1)$. Let $\underline{z}_{H_i}^{**}$ be the immediate type 0 ancestor of $\underline{z}_{H_i}$. The distribution of $\underline{z}_{H_i}$ given $\underline{z}_{H_i}^{**}, \underline{\underline{x}}, \underline{x}_0, \underline{t}$ and F is independent of all parts of the tree 'connected' to $\underline{z}_{H_i}$ only through $\underline{z}_{H_i}^{**}$. The distribution of each $z_{H_i}^{(q)}$ is Normal with mean linear in $z_{H_i}^{(q)}{}^{**}$ and $x^{(q)}$ and variance depending only on $\underline{t}$ and F;

$$\text{say } z_{H_i}^{(q)} \text{ is } N(a_i(t)z_{H_i}^{(q)}{}^{**} + b_i(\underline{t}, \underline{x}^{(q)}), V_i(t)). \qquad (4.2.9)$$

If $\underline{z}_{H_i}$ is some $\underline{x}_j$ (i.e. $n \leq i \leq (2n-1)$), $a_i(t) = V_i(t) = 0$, $b_i(\underline{t}, \underline{x}^{(q)}) = x_j^{(q)}$.

We now consider each dimension separately, and for convenience

64

drop the superscript (q).

Then

$$E(z_{H_i} | \underline{x}, x_0, \underline{t}, F) = E(E(z_{H_i} | z_{H_i}^{**}, \underline{x}, x_0, \underline{t}, F) | \underline{x}, x_0, \underline{t}, F)$$
$$= a_i(t)E(z_{H_i}^{**} | \underline{x}, x_0, \underline{t}, F) + b_i(\underline{t}, \underline{x}), \qquad (4.2.10)$$

and

$$E(z_{H_i}^2 | \underline{x}, x_0, \underline{t}, F) = E(E(z_{H_i}^2 | z_{H_i}^{**}, \underline{x}, x_0, \underline{t}, F) | \underline{x}, x_0, \underline{t}, F)$$
$$= V_i(t) + a_i(t)^2 E(z_{H_i}^{**2} | \underline{x}, x_0, \underline{t}, F) + b_i(\underline{t}, \underline{x})^2$$
$$+ 2a_i(t)b_i(\underline{t}, \underline{x})E(z_{H_i}^{**} | \underline{x}, x_0, \underline{t}, F), \qquad (4.2.11)$$

and

$$E((z_{H_i} - z_{H_i}^{**})^2 | \underline{x}, x_0, \underline{t}, F)$$
$$= E(V_i(\underline{t}) + ((a_i(\underline{t}) - 1)z_{H_i}^{**} + b_i(\underline{t}, \underline{x}))^2 | \underline{x}, x_0, \underline{t}, F)$$
$$= V_i(\underline{t}) + (a_i(\underline{t}) - 1)^2 E(z_{H_i}^{**2} | \underline{x}, x_0, \underline{t}, F)$$
$$+ 2(a_i(\underline{t}) - 1)b_i(\underline{t}, \underline{x})E(z_{H_i}^{**} | \underline{x}, x_0, \underline{t}, F) + b_i(\underline{t}, \underline{x})^2. \quad (4.2.12)$$

Explicit recurrence relations for $a_i(t)$, $b_i(\underline{t}, \underline{x})$ and $V_i(t)$ are derived in section 4.4, where the above relationships are used to compute M_k and D_k $(k = 1, \dots, (n-1))$.

4.3 THE ITERATIVE METHOD

We now state and prove the results needed to construct an iterative method, all notation being as previously defined.

Theorem 1. $\quad H(\underline{x}^{(q)}, x_0^{(q)}, \underline{t}) = \sum_{k=1}^{(n-1)} (D_k^{(q)} / t_k), \qquad (4.3.1)$

where H is as defined by (4.2.2) and $\sigma^2 = 1$.

This theorem is needed to give a simple method of evaluating the support S, but is not essential for the construction of the method. The proof, which is lengthy but straightforward, is therefore deferred to the end of the chapter (4.7).

Corollary. $-2S(\underline{x}_0, \underline{t}, F) = p\log|T| + \sum_{q=1}^{p}\sum_{k=1}^{(n-1)}(D_k^{(q)}/t_k)$

$$= p\log|T| + \sum_{k=1}^{(n-1)}(D_k/t_k). \qquad (4.3.2)$$

Proof. Rewrite equation (4.2.3) using (4.3.1) and (4.2.7). //

Lemma. $L(\underline{x}_0, \underline{t}, F) = (2\pi)^{-\frac{1}{2}K}\int\int\ldots\int_{\underline{y}}[\prod_{k=1}^{(n-1}t_k^{-\frac{1}{2}n_k p}]$

$$\exp(-\tfrac{1}{2}\sum_{k=1}^{(n-1)}C_k/t_k)d\underline{y}, \qquad (4.3.3)$$

where $K = p\sum_{k=1}^{(n-1)}n_k = \frac{1}{2}p(n+2)(n-1)$, and integration is over

$-\infty < y_i^{(q)} < \infty$, $q = 1, \ldots, p$ and $i = 1, \ldots, r_0$.

Proof. Let $\underline{z} = \{\underline{z}^{(q)}, q = 1, \ldots, p\}$ be the total set of node and population coordinates as defined above. Note that $pm = p(n+r_0) = K$.
Then

$$f(\underline{x}, \underline{y}|\underline{x}_0, \underline{t}, F) = f(\underline{z}|\underline{x}_0, \underline{t}, F)$$

$$= \prod_{j=1}^{m} f(\underline{z}_j|\underline{x}_0, \underline{t}, F, \underline{z}_i, i = 1, \ldots, (j-1))$$

$$= \prod_{j=1}^{m} f(\underline{z}_j|\underline{z}_j^*, \underline{t}, F)$$

by the independence of the separate arcs and the ordering of the $\underline{z}_i$. But
$z_j^{(q)}$ is $N(z_j^{(q)}*, t_{j*j})$, where t_{j*j} is the time length of arc $(\underline{z}_{-j}^*, \underline{z}_j)$,
and the separate dimensions q are independent. Thus

$$f(\underline{x}, \underline{y}|\underline{x}_0, \underline{t}, F)$$

$$= \prod_{j=1}^{m}[(2\pi t_{j*j})^{-\frac{1}{2}p}\exp(-\tfrac{1}{2}(\sum_{q=1}^{p}(z_j^{(q)}* - z_j^{(q)})^2/t_{j*j}))]$$

$$= (2\pi)^{-\frac{1}{2}K}[\prod_{j=1}^{m}(t_{j*j})^{-\frac{1}{2}p}]\exp(-\tfrac{1}{2}(\sum_{j=1}^{m}\sum_{q=1}^{p}(z_j^{(q)}* - z_j^{(q)})^2/t_{j*j}))$$

But now $\sum_{j=1}^{m}$ is the sum over all arcs of the tree. Hence

$$\sum_{j=1}^{m}[\,.\,] = \sum_{k=1}^{(n-1)}\sum_i\sum_j\epsilon_{ij}(k)[\,.\,].$$

Further there are n_k arcs of the tree having $t_{j*j} = t_k$. Hence, using (4.2.6),

$$f(\underline{x}, \underline{y} \,|\, \underline{x}_0, \, \underline{t}, \, F) = (2\pi)^{-\frac{1}{2}K} \prod_{k=1}^{(n-1)} (t_k^{-\frac{1}{2}p})^{n_k}] \exp(-\frac{1}{2} \sum_{k=1}^{(n-1)} \sum_{q=1}^{p} C_k^{(q)} / t_k)$$

$$= (2\pi)^{-\frac{1}{2}K} [\prod_{k=1}^{(n-1)} t_k^{-\frac{1}{2}n_k p}] \exp(-\frac{1}{2} \sum_{k=1}^{(n-1)} C_k / t_k). \qquad (4.3.4)$$

The result follows from (4.2.4). //

Theorem 2. Fundamental maximisation theorem:

$$-2 \frac{\delta S}{\delta t_k} = (n_k p / t_k) - (1/t_k)^2 E(C_k \,|\, \underline{x}, \, \underline{x}_0, \, \underline{t}, \, F)$$

$$= (n_k p / t_k^2)[t_k - M_k / n_k p]. \qquad (4.3.5)$$

[Note that $M_k / n_k p$ is the mean divergence per population per dimension during time interval t_k, given the data and the parameters. A form of (4.3.5) holds under a variety of situations (see 5.3, 5.4).]

Proof. Using (4.3.4),

$$\frac{\delta f(\underline{x}, \underline{y} \,|\, \underline{x}_0, \, \underline{t}, \, F)}{\delta t_k} = (2\pi)^{-\frac{1}{2}K} [(-\frac{1}{2} n_k p / t_k)[\prod_{j=1}^{m} t_j^{-\frac{1}{2} n_j p}] \exp(-\frac{1}{2} \sum_{j=1}^{(n-1)} C_j / t_j)$$

$$+ \frac{1}{2} (C_k / t_k^2)[\prod_{j=1}^{m} t_j^{-\frac{1}{2} n_j p}] \exp(-\frac{1}{2} \sum_{j=1}^{(n-1)} C_j / t_j)]$$

$$= (n_k p / 2 t_k^2) f(\underline{x}, \underline{y} \,|\, \underline{x}_0, \, \underline{t}, \, F)[(C_k / n_k p) - t_k].$$

Thus

$$\frac{\delta L}{\delta t_k}(\underline{x}_0, \, \underline{t}, \, F) = \int\int\cdots\int_{\underline{y}} [\frac{\delta f(\underline{x}, \underline{y} \,|\, \underline{x}_0, \, \underline{t}, \, F)}{\delta t_k}] d\underline{y} \quad \text{from (4.2.4)}$$

$$= (n_k p / 2 t_k^2) \int\int\cdots\int_{\underline{y}} [(C_k / n_k p) - t_k] f(\underline{x}, \underline{y} \,|\, \underline{x}_0, \, \underline{t}, \, F) d\underline{y}$$

$$= (n_k p / 2 t_k^2)[\int\cdots\int_{\underline{y}} (C_k / n_k p) f(\underline{y} \,|\, \underline{x}, \, \underline{x}_0, \, \underline{t}, \, F) d\underline{y} - t_k] L(\underline{x}_0, \, \underline{t}, \, F)$$

$$\text{from (4.2.4) and (4.2.5)},$$

$$= (n_k p / 2 t_k^2)[(M_k / n_k p) - t_k] L(\underline{x}_0, \, \underline{t}, \, F) \quad \text{from (4.2.8)}.$$

Thus

$$-2 \frac{\delta S}{\delta t_k} = -2 \frac{\delta (\log L)}{\delta t_k} = -(2/L) \frac{\delta L}{\delta t_k}$$

$$= (n_k p / t_k^2)[t_k - (M_k/n_k p)]$$

as required. //

Lemma. For each value of $\underline{t}$ the maximum likelihood estimate $\hat{x}_0(t)$ of $\underline{x}_0$ is given by $\hat{x}_0^{(q)}(\underline{t}) = (\underline{x}^{(q)}{}'T^{-1}\underline{1})/(\underline{1}'T^{-1}\underline{1})$, $(q = 1, \ldots, p)$, where $\underline{1}$ is a column vector of ones. (4.3.6)

Proof. Using the form of S given by (4.2.3)

$$-2S(\underline{x}_0, \underline{t}, F) = p \log |T| + \sum_{q=1}^{p} (\underline{x}^{(q)} - x_0^{(q)}\underline{1})'T^{-1}(\underline{x}^{(q)} - x_0^{(q)}\underline{1}).$$

Then $-2\dfrac{\delta S}{\delta x_0^{(q)}} = 2(x_0^{(q)}\underline{1} - \underline{x}^{(q)})'T^{-1}\underline{1}$ (T is symmetric), and

$$-2\frac{\delta^2 S}{\delta x_0^{(q)} \delta x_0^{(q*)}} = 2\delta_{q*q}(\underline{1}'T^{-1}\underline{1}), \text{ where } \delta_{q*q} = 1 \text{ if } q = q*$$
$$\text{and } \delta_{q*q} = 0 \text{ otherwise,} \quad (4.3.7)$$

whence the result follows trivially since T is a positive definite matrix. //

This lemma and Theorem 2 now give the results needed to construct an iterative method as follows:

(a)   Take some initial value of $\underline{t}$.

(b)   Find $\hat{\underline{x}}_0(\underline{t}) = \underline{\underline{x}}'T^{-1}\underline{1}/(\underline{1}'T^{-1}\underline{1})$ where $\underline{\underline{x}}' = (x_i^{(q)})$ a $p \times n$ matrix.

(c)   Find $M_k = E(C_k|\underline{\underline{x}}, \hat{\underline{x}}_0, \underline{t}, F) \quad k = 1, \ldots, (n-1)$.

(d)   Set $t_k' = M_k/n_k p \qquad\qquad k = 1, \ldots, (n-1)$.

(e)   Test for convergence of $\underline{t}$; $|t_i - t_i'|$ small, $i = 1, \ldots, (n-1)$.

(f) $\Bigg\{$ If not converged set $t_k = t_k'$ $(k = 1, \ldots, (n-1))$ and GO TO (b).

   If converged, evaluate the support $S(\underline{x}_0, \underline{t}, F)$ using (4.3.2).

Clearly if this scheme converges then the converged value is a root of the equations for a stationary value of S, within the given form F. The problems of existence, uniqueness and convergence remain. These are considered in section 4.5.

4.4   COMPUTATIONAL ASPECTS

We consider first the computational feasibility of the iterative method constructed above. We shall require the formulae of 4.2 and 4.3 and the matrix formulae given in section 4.7. To minimise computing requirements we wish only to consider type 0 nodes. Suppose first that for some given $t \; \hat{x}_0^{(q)}(t)$, $q = 1, \ldots, p$, and also the functions $a_i(\underline{t})$, $b_i(\underline{t}, \underline{x}^{(q)})$ $(q = 1, \ldots, p)$ and $V_i(\underline{t})$, $i = 1, \ldots, (n-1)$, giving the relationships (4.2.9) between type 0 nodes, are known. Then working down the tree from root to populations $E(z_{H_i}^{(q)} | \underline{x}, \underline{x}_0, \underline{t}, F)$ $(q = 1, \ldots, p)$ and $\sum_{q=1}^{p} E(z_{H_i}^{(q)2} | \underline{x}, \underline{x}_0, \underline{t}, F)$, $(i = 1, \ldots, (n-1))$, may be found using (4.2.10) and (4.2.11).

Now define the 'level' of a node to be the number of time intervals back from the present at which it occurs. Time interval t_k covers the time from level k to level (k - 1). Suppose that on a given arc the (type 1 or type 0) nodes at levels (k - 1) and k are $\underline{z}_j$ and $\underline{z}_j^*$, and that the type 0 nodes at the ends of this arc are $\underline{z}_i$ and $\underline{z}_i^{**}$ (Fig. 4.4(a)).

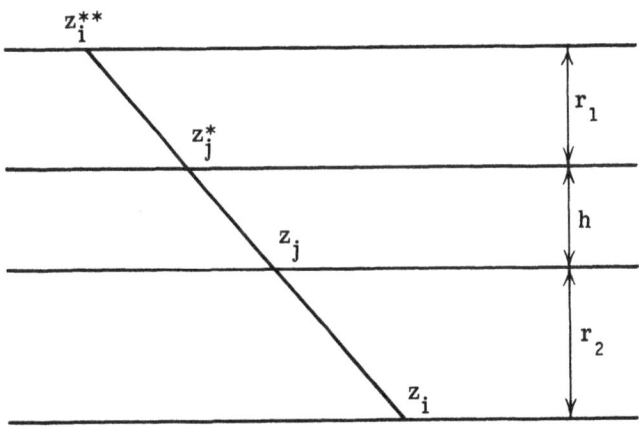

Fig. 4.4(a). The elimination of type 1 nodes; for details see text.

Thus $\underline{z}_i$, $\underline{z}_i^{**}$ are actual splitting points with level $(\underline{z}_i^{**}) \geq k$ and level $(\underline{z}_i) \leq (k - 1)$. Let the time intervals be as shown in Fig. 4.4(a). [If $\underline{z}_j$ is type 0, $\underline{z}_j \equiv \underline{z}_i$ and $r_2 = 0$. If $\underline{z}_j^*$ is type 0, $\underline{z}_j^* \equiv \underline{z}_i^{**}$ and $r_1 = 0$. Also i is some H_l, $l \leq 1 \leq 2n - 1$.] Then, again considering each dimension separately and dropping the superscript (q),

$$f(z_j, z_j^* | z_i, z_i^{**}, r_1, r_2, h)$$

$$\propto \exp[-\tfrac{1}{2}((z_i^{**} - z_j^*)^2/r_1 + (z_j - z_j^*)^2/h + (z_i - z_j)^2/r_2)]. \quad (4.4.1)$$

Hence (z_j, z_j^*) is bivariate Normal given (z_i, z_i^{**}) and

$$E((z_j - z_j^*)|z_i, z_i^{**}, r_1, r_2, h) = h(z_i - z_i^{**})/(r_1 + r_2 + h), \quad (4.4.2)$$

$$var((z_j - z_j^*)|z_i, z_i^{**}, r_1, r_2, h) = h(r_1 + r_2)/(r_1 + r_2 + h). \quad (4.4.3)$$

Thus

$$E((z_j - z_j^*)^2 | \underline{x}, \underline{x}_0, \underline{t}, F)$$

$$= E(E((z_j - z_j^*)^2 | z_i, z_i^{**}, r_1, h, \underline{x}, \underline{x}_0, \underline{t}, F) | \underline{x}, \underline{x}_0, \underline{t}, F)$$

$$= h(r_1 + r_2)/(r_1 + r_2 + h) + [h/(r_1 + r_2 + h)]^2 E((z_i - z_i^{**})^2 | \underline{x}, \underline{x}_0, \underline{t}, F).$$

$$(4.4.4)$$

Then $M_k = \sum\limits_{q=1}^{p} \sum_j$ at level $(k - 1)$ $E((z_j^{(q)} - z_j^{(q)*})^2 | \underline{x}, \underline{x}_0, \underline{t}, F)$, and

using (4.2.12) and (4.4.4) M_k may be rapidly computed from

$\sum\limits_{q=1}^{p} E(z_{H_1}^{(q)2} | \underline{x}, \underline{x}_0, \underline{t}, F)$ and $E(z_{H_1}^{(q)} | \underline{x}, \underline{x}_0, \underline{t}, F)$ $(q = 1, \ldots, p)$, if

$a_l(t)$, $b_l(t, \underline{x}^{(q)})$ $(q = 1, \ldots, p)$ and $V_l(t)$ are known $(l = 1, \ldots, (n-1))$.

To find these functions consider the subtree with root at $\underline{y}_i$ and suppose that $\underline{y}_j$ is the immediate type 0 ancestor of $\underline{y}_i$ (Fig. 4.4(b)). W.l.o.g. $\underline{y}_i = \underline{z}_{H_i}$ and $\underline{y}_j = \underline{z}_{H_i}^{**}$ for some i, $1 \le i \le (n-1)$, the ordering of the labels l of vectors $\underline{y}_l$ being immaterial (4.2). Then for each q, with the notation of Fig. 4.4(b),

$$f(y_i^{(q)} | y_j^{(q)}, \underline{x}, \underline{x}_0, \underline{t}, F)$$

$$\propto \exp[-\tfrac{1}{2}((y_i^{(q)} - y_j^{(q)})^2/h + (\underline{x}^{(q)} - y_i^{(q)}\underline{1})'S_i^{-1}(\underline{x}^{(q)} - y_i^{(q)}\underline{1}))]$$

$$\propto \exp[-\tfrac{1}{2}(y_i^{(q)2}(\underline{1}'S_i^{-1}\underline{1} + 1/h) - 2y_i^{(q)}(y_j^{(q)}/h + \underline{x}^{(q)}'S_i^{-1}\underline{1}))], \quad (4.4.5)$$

where $\underline{1}$ is the column vector of r ones, S_i the $r \times r$ covariance matrix for the subtree, and $\underline{x}^{(q)}$ the r-dimensional subtree position vector.

Hence if $U_i(t) = 1/(1 + h\underline{1}'S_i^{-1}\underline{1})$,

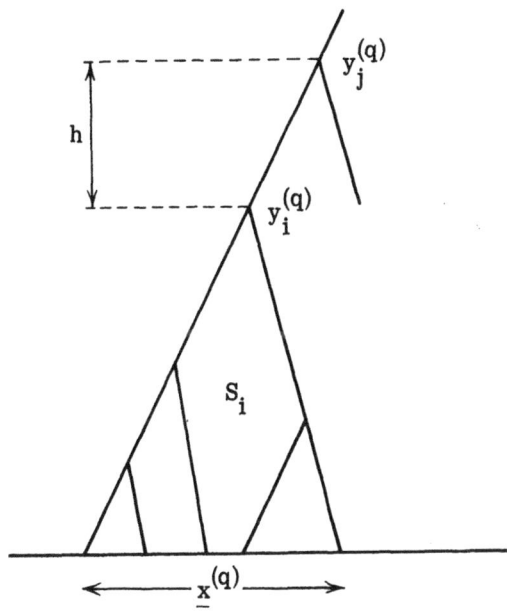

Fig. 4.4(b). The derivation of the iterative formulae; for further details see text. S_i is the r by r covariance matrix of the r-population subtree with root at $\underline{y}_i$ and $\underline{y}_j$ is the immediate type 0 ancestor of $\underline{y}_i$. $\underline{\underline{x}}$ is here the set of position vectors for the r populations of the subtree; $\underline{\underline{x}} = \{\underline{x}^{(q)}, q=1, \ldots, p\}$ and $\underline{x}^{(q)} = (\underline{x}_1^{(q)}, \ldots, \underline{x}_r^{(q)})$.

$$V_i(\underline{t}) = hU_i(\underline{t}), \quad a_i(\underline{t}) = U_i(\underline{t}) \quad \text{and} \quad b_i(\underline{t}, \underline{x}^{(q)}) = V_i(\underline{t})(\underline{x}^{(q)} \, {}'S_i^{-1}\underline{1}),$$

by comparison of the form (4.4.5) with that of (4.2.9). But the covariance matrix for the descendants of $\underline{y}_i$ is of the form;

$$S_i = \begin{pmatrix} R_i + t\underline{\underline{1}} & 0 \\ 0 & R_i^* + t^*\underline{\underline{1}} \end{pmatrix}, \tag{4.4.6}$$

where R_i and R_i^* are the covariance matrices for the subtrees having as roots the two immediate (type 0 or population) descendants of $\underline{y}_i$, t and t* are the time intervals between $\underline{y}_i$ and these two nodes, and $\underline{\underline{1}}$

is a matrix of ones of appropriate size. [The covariance matrix for a 'tree' whose root is its single population is the single element (0).] Then using the matrix formulae $(4.7.10)$ and $(4.7.11)$ we can work up the tree from populations to root evaluating, in the above notation, $\underline{1}'S_i^{-1}\underline{1}$ and $\underline{x}^{(q)}S_i^{-1}\underline{1}$, and hence $U_i(t)$, $a_i(\underline{t})$, $V_i(\underline{t})$ and $b_i(\underline{t}, \underline{x}^{(q)})$ for each type 0 node. This procedure gives also $\hat{\underline{x}}_0(\underline{\underline{x}}, \underline{t}, F)$ and, if required, $|T|$. Then working down the tree we may find M_k as described above, and hence iterate for the times.

To compute the final support, $\hat{S}$, we make use of $(4.3.2)$;

$$-2S(\underline{x}_0, \underline{t}, F) = p \log |T| + \sum_{k=1}^{(n-1)} (D_k/t_k).$$

Alternatively Felsenstein's (1973) method may be used to evaluate S, but since the mean positions of type 0 internal nodes have already been found D_k may be rapidly computed from $(4.4.2)$ and $(4.2.7)$, and $|T|$ may be computed iteratively from $(4.7.8)$ and $(4.7.11)$. Thus use of $(4.3.2)$ is more efficient in this case. Thus we see that, provided convergence is assured, the iterative method is computationally feasible. No matrix inversion or direct determinant evaluation is ever required. Evaluation of S at points in the neighbourhood of the maximum enables estimates of the curvature, and hence local two-unit support limits, to be made.

4.5   THEORETICAL ASPECTS OF THE ITERATIVE METHOD

We consider now the existence and uniqueness of stationary points of $(4.1.1)$, and the convergence of the iterative procedure of 4.3. From $(4.1.1)$,

$$-2S(\underline{x}_0, \underline{t}, F) = p \log |T| + \sum_{q=1}^{p} (\underline{x}^{(q)} - x_0^{(q)}\underline{1})'T^{-1}(\underline{x}^{(q)} - x_0^{(q)}\underline{1})$$

$$= p \log |T| + \sum_{k=1}^{n-1} D_k(\underline{\underline{x}}, \underline{x}_0, \underline{t}, F)/t_k \quad \text{(from (4.3.2))},$$

and from $(4.3.6)$ the ML estimate $\hat{x}_0(t)$, implicitly a function also of $\underline{\underline{x}}$ and F is $\hat{\underline{x}}_0(\underline{\underline{x}}, \underline{t}, F) = ((\underline{x}^{(q)}{}'T^{-1}\underline{1})/(\underline{1}'T^{-1}\underline{1}), q = 1, \ldots, p)$. Thus the maximum relative support (MRS) satisfies

$$-2S^*(\underline{t}, F) = -2S(\hat{\underline{x}}_0(\underline{x}, \underline{t}, F), \underline{t}, F)$$

$$= p\log|T| + \sum_{q=1}^{p} [\underline{x}^{(q)},T^{-1}\underline{x}^{(q)} - (\underline{x}^{(q)},T^{-1}\underline{1})^2/(\underline{1}'T^{-1}\underline{1})] \quad (4.5.1)$$

$$= p\log|T| + \sum_{k=1}^{n-1} [D_k^*/t_k] \quad \text{from } (4.3.2), \quad\quad (4.5.2)$$

where $D_k^* = D_k(\underline{x}, \hat{\underline{x}}_0(\underline{x}, \underline{t}, F), \underline{t}, F)$.

For given data $\hat{\underline{x}}_0(\underline{x}, \underline{t}, F)$ is bounded as $\underline{t}$ varies, and for each F. Hence also D_k^* $(k = 1, \ldots, (n-1))$ are bounded (see (4.2.7)). M_k is of order $\underline{t}$, but is bounded as a function of any one component of $\underline{t}$, the others, and $\underline{x}_0$, remaining fixed: so also is M_k^* $(M_k^* = M_k(\underline{x}, \hat{\underline{x}}_0(\underline{x}, \underline{t}, F), \underline{t}, F))$.

(i) The boundary conditions and the second derivatives

Unless the support surface is well-behaved at the boundaries of the parameter space, no likelihood inferences will be possible. The support function is quadratic in $\underline{x}_0$, the matrix T^{-1} being positive definite; it is sufficient to consider the MRS. From (4.3.5), the solutions to the set of equations

$$t_k = M_k^*/n_k p, \quad k = 1, \ldots, (n-1) \quad\quad (4.5.3)$$

with $t_k > 0$ are stationary points of (4.5.1) and the values of $\underline{t}$ at stationary points of (4.1.1).

$$[\frac{\delta S^*}{\delta t_k} = \frac{\delta S}{\delta t_k}\Big|_{\underline{x}_0 = \hat{\underline{x}}_0(\underline{t})} \quad \text{since} \quad \frac{\delta S}{\delta x_0^{(q)}}\Big|_{\underline{x}_0 = \hat{\underline{x}}_0(\underline{t})} = 0.]$$

$t_i = 0$ is always, for $i \neq 1$, a root of $t_i = M_i^*/n_i p$, but not necessarily of $\frac{\delta S^*}{\delta t_i} = -n_i p(t_i - M_i^*/n_i p)/2t_i^2 = 0$.

Further $t_1 = \sum_{i=1}^{n} \|\underline{x}_i - \bar{\underline{x}}\|^2/np$, where $\bar{\underline{x}} = (\sum_{i=1}^{n} \underline{x}_i)/n$, and $t_i = 0$ for $i \neq 1$, provides a solution of (4.5.3), which thus always has at least one solution. Again this solution need not be a stationary point of (4.5.1); this can only be positively asserted if $t_i > 0$ for all i. Differentiating (4.5.1) the equations $\frac{\delta S^*}{\delta t_i} = 0$ may be expressed as polynomial equations in any given t_k (the other t_i remaining fixed), the maximal degree of the polynomials depending only of n. There can thus, for each k, be at most a fixed finite number of roots of (4.5.3), and, since no degeneracy can occur, only a finite number of local maxima of S^*.

Further, as any subset of the $t_i \to \infty$, $|T| \to \infty$ while D_i^* converges to a finite limit. Thus $S^* \to -\infty$. Thus also, since $\frac{\delta S^*}{\delta t_k} = 0$ only finitely often, $-2\frac{\delta S^*}{\delta t_k} > 0$ for sufficiently large t_k (for each k, the other t_i remaining fixed), or $t_k > M_k^*/n_k p$ for all sufficiently large t_k. (4.5.4) As $t_1 \to 0$ $\log|T| \propto \log t_1$ and $D_1^*/t_1 \geq \frac{1}{2}\|\underline{x}_i - \underline{x}_j\|^2/t_1$ where $l(i, j) = 1$ [that is, $\underline{x}_i$ and $\underline{x}_j$ are the populations with splitting time s_1, see (3.1.1)], thus $-2S^* \to \infty$ and $S^* \to -\infty$ from (4.5.2). As any set of $t_i \to 0$, $(i \neq 1)$, $|T|$ converges to a finite non-zero limit, D_i^* is of order t_i^2, and S^* converges to a finite limit.

From (4.3.5) and (4.3.7) we have

$$-2S = p\log|T| + \sum_{q=1}^{p} (\underline{x}^{(q)} - \underline{x}_0^{(q)}\underline{1})'T^{-1}(\underline{x}^{(q)} - \underline{x}_0^{(q)}\underline{1})$$

$$-2\frac{\delta S}{\delta x_0^{(q)}} = -2(\underline{x}^{(q)} - \underline{x}_0^{(q)}\underline{1})'T^{-1}\underline{1}, \quad -2\frac{\delta S}{\delta t_k} = (n_k p/t_k^2)[t_k - M_k/n_k p]$$

and

$$-2\frac{\delta^2 S}{\delta x_0^{(q)}\delta x_0^{(q*)}} = 2(\underline{1}'T^{-1}\underline{1})\delta_{q*q} \quad \begin{array}{l}\text{(a positive definite} \\ \text{diagonal matrix).}\end{array} \qquad (4.5.5)$$

Using the integral expression (4.3.3) for $L(\underline{x}_0, \underline{t}, F)$, and

(i) $\quad -2\frac{\delta S}{\delta\theta} = -(2/L)\frac{\delta L}{\delta\theta}$ for any parameter θ.

(ii) $\quad -2\frac{\delta^2 S}{\delta\theta\delta\phi} = (2/L^2)[\frac{\delta L}{\delta\theta}\frac{\delta L}{\delta\phi}] - (2/L)[\frac{\delta^2 L}{\delta\theta\delta\phi}]$ for any θ, ϕ.

$$= -(2/L)\frac{\delta^2 L}{\delta\theta\delta\phi}$$ at any stationary point.

(iii) $\quad t_k = M_k/n_k p$ at any stationary point.

(iv) $\quad \int\ldots\int_{\underline{y}} h(\underline{y})f(\underline{x}, \underline{y}|\underline{x}_0, \underline{t}, F)d\underline{y}$

$$= L(\underline{x}_0, \underline{t}, F)E(h(\underline{y})|\underline{x}, \underline{x}_0, \underline{t}, F) \quad \text{(see (4.2.5))}$$

we find that, at any stationary point $(\hat{\underline{x}}_0, \hat{\underline{t}})$,

$$-2\frac{\delta^2 S}{\delta x_0^{(q)}\delta t_k} = -2[\text{cov}((y_1^{(q)} + y_2^{(q)}), C_k^{(q)})/t_{n-1}t_k^2]|_{\hat{\underline{x}}_0, \hat{\underline{t}}}, \quad (4.5.6)$$

where $\underline{y}_1$ and $\underline{y}_2$ are the immediate descendants of $\underline{x}_0$ at level $(n - 2)$, and

74

$$-2\frac{\delta^2 S}{\delta t_k \delta t_l} = [(pn_k/t_k^2)\delta_{lk} - \sum_{q=1}^{p} (\text{cov}(C_k^{(q)}, C_l^{(q)})/2t_k^2 t_l^2)]|_{\hat{\underline{x}}_0, \hat{\underline{t}}} \quad (4.5.7)$$

where all expectations are w. r. t. the conditional distribution given $\underline{\underline{x}}$ and the parameters.

For $n = 3$ it may be shown directly that the matrix of second derivatives of $-2S$ is positive definite at any stationary point, and hence, as is already known, that any stationary point is a local maximum of the support function, and hence unique. The covariances are not zero at the stationary points, and in general it may not be true that all stationary points are local maxima.

(ii) **Existence of internal roots** $(\hat{t}_i > 0$ for all i) **and change of tree form**

We have seen (3. 3(ii)) that for $n = 3$ there are cases when there is no tree form with a maximum in $t_2 > 0$, and that this may occur over a large range of population positions in cases that are in no way pathological. Thus it may be that for every F the tree of maximum support may be non-bifurcating. Alternatively there may be several tree forms with maxima in $t_i > 0$ for all i. We know that for $n = 3$ this cannot occur, and further that, if it exists, the ML tree form is the unique form having an internal root. We would like to assert some general hypothesis along these lines but this does not seem to be possible. However a tree form with zero ML estimate, $\hat{t}_k$, must have a support no greater than that of any tree form obtained by changing the labelled history about interval t_k, since the two forms have equal support at $t_k = 0$. The support for the new form will be strictly greater if this form has an internal maximum. This provides a criterion for change of tree form.

The iterative method so far presented gives only a method of finding the ML estimates of $\underline{t}$ and $\underline{x}_0$ within any given F. If for some given F t_k converges to zero, we will obtain a tree of at least as great a support by changing the tree form about t_k, and continuing iteration with this new form. Thus the problem of which tree forms should be considered reduces to finding an initial tree at which to start iteration. There are two cases of 'changing the form about interval t_k', one of which involves the testing of two alternative forms (Fig. 4. 5(a)). Iteration is

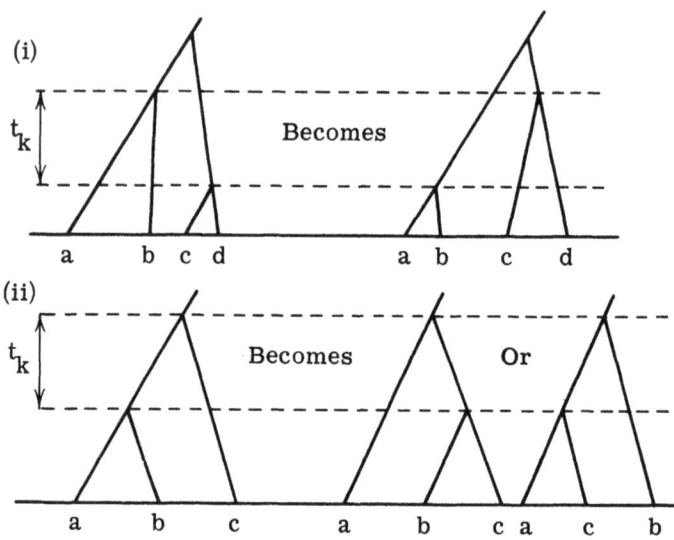

(i)

t_k

Becomes

a   b c d   a b   c   d

(ii)

t_k

Becomes   Or

a   b   c   a   b   c a   c   b

Fig. 4.5(a).   The possible changes of tree form about t_k when
$\hat{t}_k = 0$.   (i) Change of history only.   (ii) Change
of topology.

continued until either we have an F with an internal maximum, or until
no changes about any zero t_k produces an increase in support, the alter-
natives also having $\hat{t}_k = 0$.

If there are many tree forms with internal maxima we may cease
iteration before reaching the ML form. This may also occur if there is
a local maximum for which $\hat{t}_k = 0$ for a tree and all its immediate alter-
natives, although some other change through two steps may improve the
support. In practice we may investigate whether such local maxima exist
by starting iteration from different F. In fact at most one form with an
internal maximum has been found for any one data set, and more often,
for large n, there is no such form and the relevant time intervals converge to zero from any chosen starting point.

Although we often obtain a non-bifurcating tree it is often only the
root that is non-bifurcating (see the examples of 5.1). Felsenstein (1973)
has suggested that the tendency of the MRL to produce non-bifurcating
roots is an indication that the MRL is inappropriate. However it is not

true that a non-bifurcating tree is always produced.

Lemma. For all n there exist $(\underline{x}_1, \ldots, \underline{x}_n)$, points in some Euclidean space (of dimension $(n-1)$) such that the ML tree has $\hat{t}_i > 0$, $i = 1, \ldots, (n-1)$.

Proof. Note that although if $p < (n-1)$ not all possible Euclidean patterns of population distances are obtainable, ML estimates of the times depend only on the pairwise population distances divided by p. The result is independent of p.

The result holds for $n = 3$; suppose true also for $n = r - 1$, the time estimates being $\hat{t}_1, \ldots, \hat{t}_{r-2}$. Then continuity and the form of M_k^* ensures that for any $\delta > 0$ there exists an $\varepsilon_r > 0$, such that if

$$\| \underline{x}_r - \underline{x}_{r-1} \| \le \varepsilon_r$$

then the r population tree with $l(r-1, r) = 1$ has time estimates t_i^* with

$$| t_i^* - \hat{t}_{i-1} | < \delta \quad (i = 2, \ldots, (r-1))$$

and

$$t_1^* > 0 \text{ if } \underline{x}_r \ne \underline{x}_{r-1}.$$

The smallest t_i^* may then be made as large as we please by scaling the population distances, and by induction the result is proved. //

We have a bifurcating root $(\hat{t}_{n-1} > 0)$ provided the populations fall 'sufficiently' into two groups, where precise but not very meaningful formulae may be given for 'sufficiently'.

A further conjecture is more interesting; namely that for all $t_1, \ldots, t_{n-1}$ (t_i non-negative) there exist $\underline{x}_1, \ldots, \underline{x}_n$ and F such that $\underline{\underline{t}}$ is the ML time estimate for $\underline{x}$ and F. This conjecture has not been rigorously proved, but it seems that, at least provided $p \ge (n-1)$, we have sufficient freedom to obtain any required pattern of covariances.

(iii) **Unimodality and convergence of the iterative method**

A more serious problem, from the point of view of likelihood inference, than the possible non-existence of tree forms with internal maxima, is the possibility of the existence of more than one stationary point within any one tree form. All stationary points of (4.1.1) are maxima w. r. t. $x_0^{(q)}$ $(q = 1, \ldots, p)$ (4.3), but $S^*(\underline{t}, F)$ may have stationary points of any type. For a unique maximum we require that the implicit equations (4.5.2) have at most one strictly positive root.

The question of convergence of the iterative method is closely connected with that of the existence of a unique maximum. The iterative method sets $t_k' = M_k^*/n_k p$ (see 4.3) and thus

$$-2 \frac{\delta S^*}{\delta t_k} = (n_k p/t_k^2)[t_k - t_k'] \qquad (4.5.8)$$

or

$$t_k \gtrless t_k' \quad \text{as} \quad \frac{\delta S^*}{\delta t_k} \lessgtr 0.$$

The method thus causes the estimate of $\underline{t}$ to 'climb the support surface', and we may expect convergence to a local maximum, although this is not theoretically necessary. (The estimate could oscillate indefinitely.) Local maxima are the only stable points of convergence of the iterative method; minima and saddle points are unstable.

Note that we have

$$t_k' = t_k + (2t_k^2/n_k p) \frac{\delta S^*}{\delta t_k},$$

whereas, in standard notation, Newton-Raphson iteration would give

$$t_k' = t_k + \sum_{l=1}^{n-1} [-\frac{\delta^2 S^*}{\delta t^2}]^{kl} \frac{\delta S^*}{\delta t_l}.$$

From the second derivatives of S ((4.5.5)-(4.5.7)) we may obtain the matrix of second derivatives of S^*, and we see that the iterative procedure corresponds to the inverse of the first term of the diagonal components of the matrix. Were it computationally feasible Newton-Raphson iteration would converge to any stationary point. Our procedure converges only to maxima: it has a larger range of convergence but is of only first order.

The function S^* has precisely one maximum along any ray

$$\{c\underline{t}^*; \ c \geq 0, \ \underline{t}^* \ \text{fixed}\}. \qquad (4.5.9)$$

For D_k and $\hat{x}_0(\underline{x}, \underline{t}, F)$ depend only on relative times, and from (4.5.2)

$$D_k^*(\underline{x}, \underline{ct}^*, F) = D_k^*(\underline{x}, \underline{t}^*, F)$$

and

$$-2S^*(\underline{ct}^*, F) = np \log c + p \log|T| + (1/c) \sum_{k=1}^{n-1} D_k^*/t_k^* ,$$

where T and D_k^* are evaluated at $\underline{x}$, $\underline{t}^*$, F.

We may further consider the support as some t_k varies, the other t_i remaining fixed. For a stationary point w. r. t. t_k we require

$$-2\frac{\delta S^*}{\delta t_k} = (n_k p/t_k^2)(t_k - M_k^*/n_k p) = 0. \tag{4.5.10}$$

M_k^* is an increasing function of t_k, and $M_k^*/n_k p < t_k$ for all sufficiently large t_k (see (4.5.4)).

As $t_k \to 0$, $M_k^* = pn_k t_k + a_k t_k^2 + 0(t_k^3)$ where a_k is independent of t_k, $(k \neq 1)$. Further from the definition of M_k (4.2.8), and using the considerations of 4.4, it may be shown that M_k^* has at most one point of inflection for varying t_k. We thus have two possible cases (Fig. 4.5(b)), and there is at most one positive root of (4.5.10).

$$a_k = \lim_{t_k \to 0} [(M_k^* - n_k p t_k)/t_k^2] = \lim_{t_k \to 0} (2\frac{\delta S^*}{\delta t_k}).$$

If $a_k > 0$ there is a unique root t_k^*. Since $M_k^* > t_k n_k p$ only if $t_k < t_k^*$, t_k^* gives a maximum of S^* (see (4.5.10)). If $a_k < 0$ there is no root of (4.5.10) in $t_k > 0$.

For $k = 1$, $M_1 \to \frac{1}{2}\|\underline{x}_i - \underline{x}_j\|^2$ as $t_1 \to 0$, where $l(i, j) = 1$, and there is always one root of (4.5.10) in $t_1 > 0$.

Thus S^* is unimodal in each t_k, and together with (4.5.9) this gives a clear idea of the possible support surfaces. Although for $n = 3$ we have overall unimodality, there seems to be no reason why in general we should not have a support surface of the form shown in Fig. 4.5(c).

We return now to the problem of convergence. We see from Fig. 4.5(b) that, considering iteration only in the kth dimension, we have convergence to t_k^*, or to zero if no such t_k^* exists. This ensures monotone convergence in the general iteration in regions in which the sets where S^* is greater than some given constant are convex, but not neces-

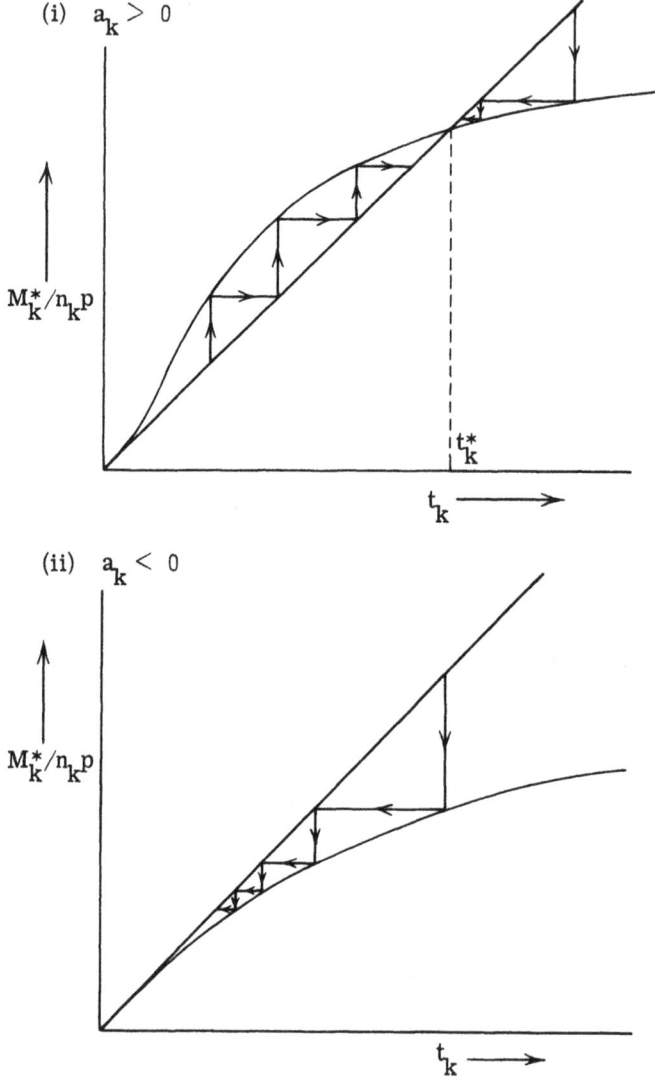

(i) $a_k > 0$

$M_k^*/n_k p$

t_k^*

$t_k \longrightarrow$

(ii) $a_k < 0$

$M_k^*/n_k p$

$t_k \longrightarrow$

Fig. 4.5(b).   The two possible forms of $M_k^*/n_k p$ as a function
of t_k, showing iteration for the root of the equation
$t_k = M_k^*/n_k p$.

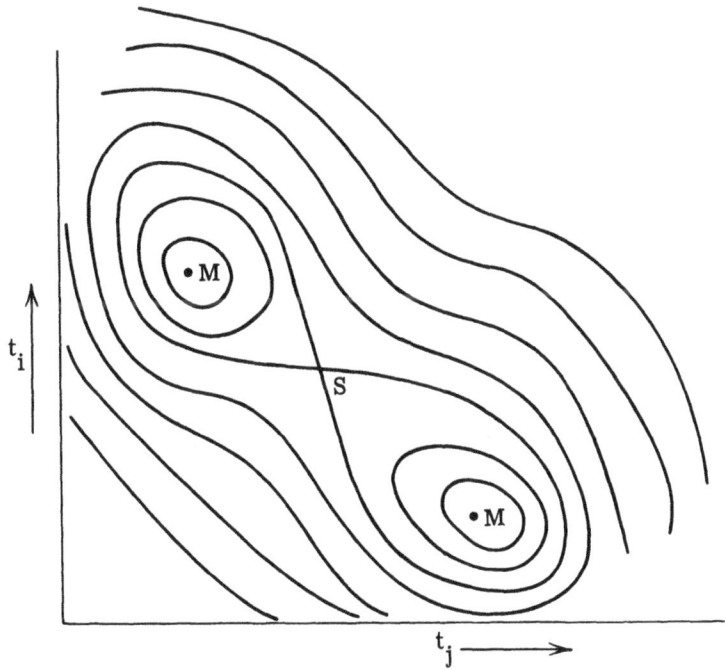

Fig. 4.5(c). Contours of a possible maximum relative support surface. Points M are local maxima, while S is a saddle point.

sarily in general. Thus although in practice we always have rapid convergence to a local maximum, we cannot assert that this is the only one or that convergence will always be obtained. We note finally that when $a_k < 0$ convergence is very slow, the difference between successive iteratives being of order t_k^2. Often when it is seen that some t_k is converging to zero it is more efficient to test immediately whether $t_k = 0$ gives higher support.

4.6   FURTHER ASPECTS OF THE LIKELIHOOD SOLUTION

(i)   Simplification of the projected coordinates

The functions of the gene frequencies that are assumed to undergo Brownian motion are, for locus i having k_i alleles, any $(k_i - 1)$ orthogonal coordinates in the projected space (2. 3). The overall space has dimension $p = \sum_{i=1}^{s} (k_i - 1)$. This approximation is valid provided that, for the loci in question, the major cause of observed gene frequency differentiation is r. g. d. ; the problem of sampling remains to be considered. We require also that population sizes are approximately equal at any one time, that all allele frequencies lie between 0. 05 and 0. 95 and that $\sum_{i=1}^{n-1} t_i/N_e \leq 0.1$, where t_i are now measured in generations. Under these restrictions discussed in Chapter 2 we may hope to make valid inferences about the form of evolutionary tree.

It complicates the problem unnecessarily to take as the data variables the actual observed projected coordinates. Brownian motion is independent of the particular coordinate system chosen, and the likelihood depends only on population distances, including distances from $\underline{x}_0$. Thus if, as is often the case, $(n-1) < p$, the number of variables which must be considered may be reduced by embedding the projected space population distances in a Euclidean space of $(n - 1)$ dimensions, as is done for the heuristic ME solution (Cavalli-Sforza and Edwards (1967)). This embedding is equivalent to a rotation and translation of the projected space in which all the coordinates

$$x_i^{(q)} \text{ for } 1 \leq i \leq n, \text{ and } n \leq q \leq p$$

have become zero. Then $\hat{x}_0^{(q)} = 0$ for $n \leq q \leq p$, and the mean internal node positions, also lie in this subspace. The dimension of the motion is still p; thus embedding need not be rejected, as it is by Malyutov et al. (1972), on the grounds that it reduces the Brownian motion dimension. The last $(p - n + 1)$ dimensions can be ignored in computation of mean positions, but still contribute to the variance terms in M_k. In programming there need now be no restriction on p, since the number of variables that must be retained is independent of p.

(ii) **Singularities and past data**

Lemma. For contemporary populations the covariance matrix T is non-singular if and only if $t_1 > 0$.

Proof. Let the two major subtrees of the tree with matrix T have covariance matrices S_1 and S_2 and let t and t^* be the times from $\underline{x}_0$ to the two immediate type 0 descendants of $\underline{x}_0$. Suppose that S_1 and S_2 are positive definite. Then from (4.4.6)

$$T = \begin{pmatrix} S_1 + t\underline{1} & 0 \\ 0 & S_2 + t^*\underline{1} \end{pmatrix}$$

and from (4.7.8)

$$|T| = |S_1|.|S_2|.(1 + t\underline{1}'S_1^{-1}\underline{1})(1 + t^*\underline{1}'S_2^{-1}\underline{1}) > 0 .$$

Thus, since T is a covariance matrix, T is positive definite. Conversely suppose that $t_1 = 0$, and let x_i and x_j be the two populations with $l(i, j) = 1$. Then $\text{cov}(x_i^{(q)}, x_r^{(q)}) = \text{cov}(x_j^{(q)}, x_r^{(q)})$ for all $r \neq i, j$, and $\text{var}(x_i^{(q)}) = \text{var}(x_j^{(q)}) = \text{cov}(x_i^{(q)}, x_j^{(q)}) = \sum_{k=2}^{n-1}(t_k)$, for each q, from (4.2.1). Thus T is singular having two identical rows. //

The restriction $t_1 > 0$ places no restriction on the iterative method. If two populations have identical coordinates they should be considered as a single population. Under any other circumstances we shall never reach an estimate $t_1 = 0$, since as $t_1 \to 0$, $S \to -\infty$; the support surface has no positive infinities.

However if the data are not contemporary, the evolution of some population(s) does not span t_1. If the data are at known times in the past we have an absolute scale of time, for example in years, and σ^2 is no longer a scale factor. Time intervals, t_k, and divergences, D_k, may be defined as before, but with additional intervals known in terms of the $(n - 1)$ splitting intervals being given by the non-contemporary population points. The form of T may be readily modified and (4.3.2) remains true. Replacing $\underline{t}$ by $\sigma^2\underline{t}$ in (4.3.2) we have $\hat{\sigma}^2 = (1/np)\sum_k(D_k/t_k)$, and we consider the MRS

$$-2S^*(\underline{x}_0, \hat{\sigma}^2, \underline{t}, F) = p\log|T| + np\log(\sum_k D_k(\underline{x}, \underline{x}_0, \underline{t}, F)/t_k) + \text{constant}. \quad (4.6.1)$$

If σ^2 is known, or if we have information on the times of past data only in terms of σ^2, the support function is of the same form as before;

$$-2S(\underline{x}_0, \underline{t}, F) = p \log|T| + \sum_k (D_k(\underline{x}, \underline{x}_0, \underline{t}, F)/t_k), \quad [\sigma^2 = 1]. \quad (4.6.2)$$

In either case suppose that it is either possible or given that there is some data point strictly previous to all the others, say $\underline{x}_r$, and consider the situation when the time previous to $\underline{x}_r$ becomes zero; $t_{i_0} \to 0$ for all $i > i_0$ say, and $\sum_{i=1}^{i_0} t_i \to$ 'time ago of $\underline{x}_r$'. Provided $\underline{x}_0 = \underline{x}_r$, D_i is $0(t_i^2)$ as $t_i \to 0$ for each $i > i_0$, and $D_i/t_i \to 0$;

$$\sum_k (D_k/t_k) \to \sum_{i \le i_0} (D_i/t_i); \quad \text{a finite limit.}$$

But if $\underline{x}_r$ has zero evolutionary time from $\underline{x}_0$,

$$\text{var}(x_r^{(q)}) = 0 = \text{cov}(x_r^{(q)}, x_j^{(q)}) \quad \text{for each } j \ne r \text{ and } q = 1, \ldots, p,$$

and $\log|T| \to -\infty$; $S(\underline{x}_0, \underline{t}, F)$ and $S^*(\underline{x}_0, \hat{\sigma}^2, \underline{t}, F) \to +\infty$. We have infinite support for the hypothesis that the point of origin is the position and time of the earliest data population, which is an unreasonable proposition.

If there are two or more distinct populations $(\underline{x}_r, i=1, \ldots, w$, $2 \le w < n$ say) at the most previous time point we do not have an infinite singularity. Suppose $l(r_1, r_2) = l > i_0$, and that the populations are distinct. As $t_i \to 0$ for all $i > i_0$, we have as above $\log|T| \to -\infty$, being of order of, at worst, $w \log(t_l)$, but $D_l \to \|\underline{x}_{r_1} - \underline{x}_0\|^2 + \|\underline{x}_{r_2} - \underline{x}_0\|^2 \ge \frac{1}{2} \|\underline{x}_{r_1} - \underline{x}_{r_2}\|^2$.

Thus $\sum_k (D_k/t_k)$ is of order $1/t_l$ and $S(\underline{x}_0, \underline{t}, F) \to -\infty$. Also $S^*(\underline{x}_n, \hat{\sigma}^2, \underline{t}, F) \to -\infty$ $(n > w)$ $[\hat{\sigma}^2 \to +\infty]$.

Thus, provided the order of magnitude of σ^2 is known, it should be possible to estimate an evolutionary tree. [If there is no prior information at all on either σ^2 or the times of ancestral nodes we may have the further anomaly of maximal, although not infinite, support for a tree of 'almost zero' divergence rate, with roots 'almost infinitely' long ago.]

The problem of non-contemporary data does not arise when trees are to be constructed on the basis of blood group frequencies, as there is no possibility of having such data for past times. However, this problem of the singularities which arise as soon as past data are admitted is an interesting likelihood problem, which could arise in practice for trees based upon anthropometric data. [Although a Normal model may not be valid for anthropometric data, the same problem arises with any similar diffusion model.] It also raises the problem of what is meant by contemporary. Although our acceptance of present-day populations as contemporary seems justifiable, in fact gene frequencies are not measured at precisely the same instant, and may be as much as a generation old.

(iii) The use of prior information

Prior information on either tree form or times of split may in theory be included by means of a prior likelihood (1.3); but this may cause problems in the iterative method. Let the prior support for F and $\underline{t}$ be $\log g(F)$ and $\log h(\underline{t})$ respectively.

Then the net support satisfies

$$-2S(\underline{x}_0, \sigma^2, \underline{t}, F) = -2 \log h(\underline{t}) - 2 \log g(F)$$
$$+ p \log |\sigma^2 T| + \sum_{k=1}^{n-1} (D_k/\sigma^2 t_k). \qquad (4.6.3)$$

The prior support for F enters only into the comparison of different tree forms, and so does not affect the iterative method explicitly. However there may now be discontinuities in the support at points of change of form $(t_k = 0)$. This is intuitively undesirable, and complicates the criteria for change of form. Any large variation in $g(F)$, over F, is likely to dominate over any presently available genetic support.

Unless $h(\underline{t})$ is a homogeneous function of degree 0 in $\underline{t}$, or is a function of $\sigma^2 t$, σ^2 is no longer a scale factor. There may often be a case for introducing some support function for relative times; this may be a better way of incorporating desirable restrictions on splitting time intervals than the inclusion of a probability model for population splitting. If $h(\underline{t})$ is only a function of relative times, or of times relative to σ^2, the iterative procedure may be adapted. At any stationary point we now have

$$t_k = (1/n_k p)[M_k(\underline{x}, \hat{\underline{x}}_0(\underline{x}, \underline{t}, F), \underline{t}, F) + 2 \frac{\delta h(\underline{t})}{\delta t_k}(t_k^2/h(\underline{t}))],$$

where $\hat{\underline{x}}_0(\underline{x}, \underline{t}, F)$ is unchanged by the inclusion of $h(\underline{t})$.

Thus the iterative procedure may be modified to

$$t_k' = (1/n_k p)[M_k^* + 2\frac{\delta h}{\delta t_k} (t_k^2/h(\underline{t}))] \qquad [\text{cf. 4.3}].$$

At each stage t_k will be changed in the direction of increasing support (cf. (4.5.8)). Provided h is a suitable function we should in practice obtain convergence to a local maximum, although the form of h may now cause there to be several of these.

4.7 APPENDICES TO CHAPTER 4

Appendix 1. Proof of Theorem 1 of section 4.3.

Theorem. $H(\underline{x}, \underline{x}_0, \underline{t}) = \sum_{k=1}^{n-1} (D_k/t_k),$ (4.7.1)

where H is defined by (4.2.1) and (4.2.2) and D_k by (4.2.7), being the value given by a single dimension of the tree, with population positions $\underline{x}$.

Proof. We proceed by induction, so note first that the result is trivially true for a tree of two populations;

$$T = \begin{pmatrix} t_1 & 0 \\ 0 & t_1 \end{pmatrix}.$$

We assume that the result is true for each of the two major subtrees $\underline{Q}_1$ and $\underline{Q}_2$ of a tree $\underline{P}$. Let $\underline{Q}_1[\underline{Q}_2]$ have covariance matrix S [R], root $z_1[z_2]$, populations $\underline{x}^{(1)}[\underline{x}^{(2)}]$ and time intervals $\underline{s}[\underline{r}]$ (Fig. 4.7(a)): The combined tree $\underline{P}$ has root x_0, populations $\underline{x}$ [$\underline{x}' = (\underline{x}^{(1)'}, \underline{x}^{(2)'})$], and time intervals $\underline{t}$.

Let $H(\underline{P})$ be the value of H for a tree $\underline{P}$, and $F(\underline{P})$ be the form of $\underline{P}$.

Let $D_k(\underline{P})$ be the value of D_k given by $\underline{P}$, or by $\underline{x}$, $\underline{x}_0$, $\underline{t}$ and $F(\underline{P})$, and $D_k(\underline{P})|_{Q_i}$ be $D_k(\underline{P})$ modified by the restriction that the sum is only over those arcs of $\underline{P}$ in $\underline{Q}_i$. //

First we require two lemmas,

Lemma 1. Let $m_i(\underline{Q}_1) = E(y_i|\underline{x}^{(1)}, z_1, \underline{s}, F(\underline{Q}_1))$ and $m_i(\underline{P}) = E(y_i|\underline{x}, \underline{x}_0, \underline{t}, F(\underline{P}))$, where y_i is any internal node of $\underline{Q}_1$.

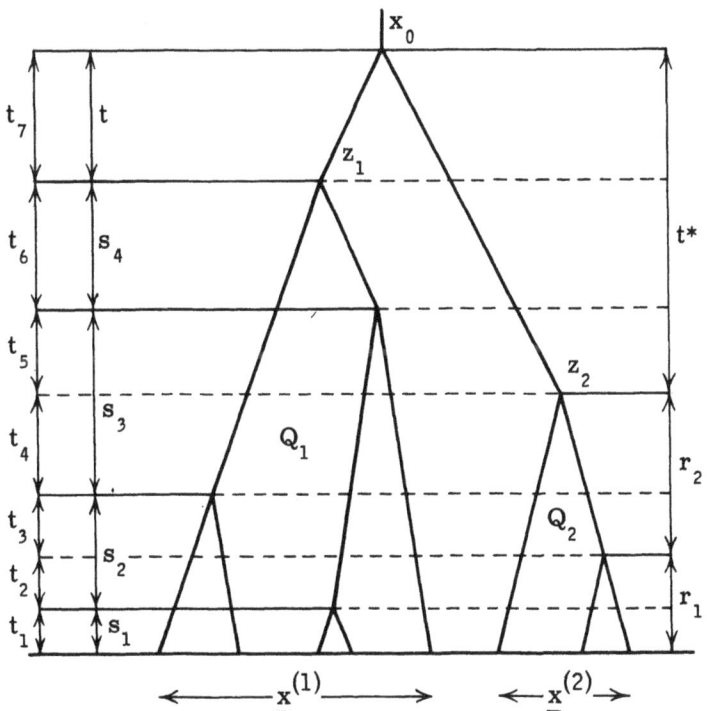

Fig. 4.7(a). Division of the one-dimensional tree $\underline{P}$ into its two major subtrees $\underline{Q}$ and $\underline{Q}_2$, as required in the proof of Theorem 1.

Then if $z_1 = E(Z_1 | \underline{x}, x_0, \underline{t}, F(\underline{P}))$, where Z_1 denotes the random variable of the position of the root of $\underline{Q}_1$ under tree $\underline{P}$,

$$m_i(\underline{Q}_1) = m_i(\underline{P}) \qquad (4.7.2)$$

with an equivalent result for the nodes of $\underline{Q}_2$.

Proof of Lemma 1. If $\underline{x}$, $\underline{t}$ and $F(\underline{P})$ are given, so automatically are $\underline{x}^{(1)}$, $\underline{s}$ and $F(\underline{Q}_1)$. The result then follows immediately from the linearity of m_i in the conditioning variables (see 4.2);

$$m_i(\underline{P}) = E(y_i | \underline{x}, x_0, \underline{t}, F(\underline{P}))$$
$$= E(E(y_i | \underline{x}^{(1)}, Z_1, F(\underline{Q}_1), \underline{s}) | \underline{x}, x_0, \underline{t}, F(\underline{P}))$$

$$= E(k_1(\underline{x}^{(1)}, \underline{s})Z_1 + k_2(\underline{x}^{(1)}, \underline{s})|\underline{x}, \underline{x}_0, \underline{t}, F(\underline{P}))$$

$$= k_1 E(Z_1|\underline{x}, \underline{x}_0, \underline{t}, F(\underline{P})) + k_2$$

$$= m_i(\underline{Q}_1) \text{ for the given } z_1. \; /\!/$$

Lemma 2. 'Type 1 nodes may be inserted at any point' or: If y_i^* and y_i are the positions of a population before and after time interval s_k and z is the mean position, given (y_i^*, y_i), of the population at a time h from the beginning of the interval (see Fig. 4.7(b)),

$$(y_i - y_i^*)^2/s_k = (y_i^* - z)^2/h + (y_i - z)^2/(s_k - h). \tag{4.7.3}$$

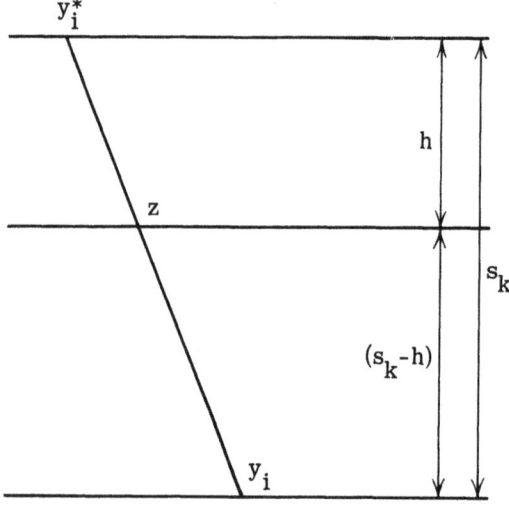

Fig. 4.7(b).   Division of an arc of the tree, as required in the proof of Lemma 2.

Proof of Lemma 2. $z = (y_i^*(s_k - h) + y_i h)/s_k.$
Clearly the result follows. $/\!/$

Corollary to the two lemmas.   If $z_1 = E(Z_1|\underline{x}, \underline{x}_0, \underline{t}, F(\underline{P}))$,

$$\Sigma_k(D_k(\underline{Q}_1)/s_k) = \Sigma_j(D_j(\underline{P})|_{\underline{Q}_1} /t_j). \tag{4.7.4}$$

Proof. Using Lemma 2 to insert type 1 nodes in $\underline{Q}_1$ at the points in time at which there are splits in $\underline{Q}_2$ (Fig. 4.7(a)), and Lemma 1 to ensure no change in mean population positions, for each j,

$$D_j(\underline{Q}_1)/s_j = \sum_{k \in A_j} (D_k(\underline{P})|_{\underline{Q}_1} / t_k),$$

where A_j is the set of k for which time interval t_k of $\underline{P}$ is part of interval s_j of $\underline{Q}_1$.

Thus

$$\sum_j D_j(\underline{Q}_1)/s_j = \sum_j \sum_{k \in A_j} (D_k(\underline{P})|_{\underline{Q}_1} / t_k) = \sum_k (D_k(\underline{P})|_{\underline{Q}_1} / t_k). \; /\!/$$

We return now to the proof of the main theorem; as in (4.4.6)

$$T = \begin{pmatrix} S + t\underline{\underline{1}} & 0 \\ 0 & R + t^*\underline{\underline{1}} \end{pmatrix},$$

where $\underline{\underline{1}}$ is a square matrix of ones of appropriate size, and t and t^* are as shown in Fig. 4.7(a). If $\underline{Q}_i$ consists of only one population S (or R) is the single element (0). Suppose for the moment that this is not the case. Then r_1 and s_1 are strictly positive and R and S are non-singular (see 4.6(ii)).

Let $w = 1 + t\underline{1}'S^{-1}\underline{1}$ and $w^* = 1 + t^*\underline{1}'R^{-1}\underline{1}$, where $\underline{1}$ is a column vector of ones of appropriate length.

From the matrix formula (4.7.9);

$$H(\underline{P}) = (\underline{x} - \underline{x}_0\underline{1})'T^{-1}(\underline{x} - \underline{x}_0\underline{1})$$
$$= (\underline{x}^{(1)} - \underline{x}_0\underline{1})'[S^{-1} - (tS^{-1}\underline{1}\underline{1}'S^{-1})/w](\underline{x}^{(1)} - \underline{x}_0\underline{1})$$
$$+ (\underline{x}^{(2)} - \underline{x}_0\underline{1})'[R^{-1} - (t^*R^{-1}\underline{1}\underline{1}'R^{-1})/w^*](\underline{x}^{(2)} - \underline{x}_0\underline{1})$$
$$= G_1 + G_2 \text{ say.} \qquad (4.7.5)$$

But $f(Z_1 | \underline{x}, x_0, \underline{t}, F(\underline{P}))$

$$\propto \exp[-\tfrac{1}{2}((Z_1 - x_0)^2 / t + (\underline{x}^{(1)} - Z_1\underline{1})'S^{-1}(\underline{x}^{(1)} - Z_1\underline{1}))]$$

and thus if $z_1 = E(Z_1 | \underline{x}, x_0, \underline{t}, F(\underline{P}))$

89

$$(z_1 - x_0)/t = \underline{1}'S^{-1}(\underline{x}^{(1)} - z_1\underline{1}). \qquad (4.7.6)$$

Then

$$
\begin{aligned}
G_1 &= (\underline{x}^{(1)} - z_1\underline{1} + (z_1 - x_0)\underline{1})'(S^{-1} - (S^{-1}\underline{1}\underline{1}'S^{-1})/w)(\underline{x}^{(1)} - z_1\underline{1} + (z_1 - x_0)\underline{1}) \\
&= (\underline{x}^{(1)} - z_1\underline{1})'S^{-1}(\underline{x}^{(1)} - z_1\underline{1}) - (t/w)((\underline{x}^{(1)} - z_1\underline{1})'S^{-1}\underline{1})^2 \\
&\quad + 2(\underline{x}^{(1)} - z_1\underline{1})'S^{-1}\underline{1}(z_1 - x_0)/w + (z_1 - x_0)^2(\underline{1}'S^{-1}\underline{1})/w \\
&= H(\underline{Q}_1) + ((z_1 - x_0)^2/tw)[-1 + 2 + t\underline{1}'S^{-1}\underline{1}] \quad \text{(using (4.7.6))}, \\
&= H(\underline{Q}_1) + (z_1 - x_0)^2/t.
\end{aligned}
$$

The requirement that $\underline{Q}_1$ does not consist of only one population may now be dropped, since in this case

$$z_1 = x, \text{ where } x \text{ is the population position,}$$

and

$$G_1 = (x - x_0)^2/t, \text{ where } t \text{ is the total time of } \underline{P},$$

and we may take $H(\underline{Q}_1) = 0$. Then from (4.7.5)

$$
\begin{aligned}
H(\underline{P}) &= H(\underline{Q}_1) + H(\underline{Q}_2) + (z_1 - x_0)^2/t + (z_2 - x_0)^2/t* \\
&\quad \text{provided } z_i = E(Z_i|\underline{x}, x_0, \underline{t}, F(\underline{P})) \\
&= \Sigma_j(D_j(\underline{Q}_1)/s_j) + \Sigma_i(D_i(\underline{Q}_2)/r_i) + (z_1 - x_0)^2/t + (z_2 - x_0)^2/t* \\
&\quad \text{by the inductive hypothesis} \\
&= \Sigma_k(D_k(\underline{P})|_{\underline{Q}_1}/t_k) + \Sigma_k(D_k(\underline{P})|_{\underline{Q}_2}/t_k) \\
&\quad + (z_1 - x_0)^2/t + (z_2 - x_0)^2/t* \quad \text{(from (4.7.4))} \\
&= \Sigma_k D_k(\underline{P})/t_k \quad \text{on again using Lemma 2 to insert a type 1} \\
&\qquad \text{node on arc } (x_0, z_2) \text{ at time } t \text{ from } x_0 \\
&\qquad \text{(w. l. o. g. } t* > t). \qquad (4.7.7)
\end{aligned}
$$

Since all trees have some two population subtree, the required result is proved by induction.

Appendix 2.   Some standard matrix formulae.

The following results are used repeatedly in proving the results of Chapter 4, and are given here for reference. They may all be verified directly.

Suppose that S and R are positive definite symmetric square matrices of ranks n_1 and n_2, and that $n = n_1 + n_2$. Let $\underline{1}$ and $\underline{1}$ denote the square matrix and column vector of ones, of any appropriate size, and $'$ denote the transpose of a vector or matrix. Let $\underline{x}^{(1)}$ and $\underline{x}^{(2)}$ be any column vectors of lengths n_1 and n_2, and $\underline{x}' = (\underline{x}^{(1)\prime}, \underline{x}^{(2)\prime})$.

Let

$$T = \begin{pmatrix} S + t_s\underline{1} & 0 \\ 0 & R + t_r\underline{1} \end{pmatrix},$$

where t_s, t_r are any given non-negative scalars.   Then

$$|T| = |S + t_s\underline{1}||R + t_r\underline{1}| = |S||R|(1 + t_s\underline{1}'S^{-1}\underline{1})(1 + t_r\underline{1}'R^{-1}\underline{1}), \quad (4.7.8)$$

and

$$T^{-1} = \begin{pmatrix} (S + t_s\underline{1})^{-1} & 0 \\ 0 & (R + t_r\underline{1})^{-1} \end{pmatrix}$$

$$= \begin{pmatrix} (S^{-1} - (t_s S^{-1}\underline{1}\,\underline{1}'S^{-1})/(1 + t_s\underline{1}'S^{-1}\underline{1})) & 0 \\ 0 & (R^{-1} - (t_r R^{-1}\underline{1}\,\underline{1}'R^{-1})/(1 + t_r\underline{1}'R^{-1}\underline{1})) \end{pmatrix}.$$

$$(4.7.9)$$

Hence

$$\underline{x}'T^{-1}\underline{1} = \underline{x}^{(1)\prime}S^{-1}\underline{1}/(1 + t_s\underline{1}'S^{-1}\underline{1})$$
$$+ \underline{x}^{(2)\prime}R^{-1}\underline{1}/(1 + t_r\underline{1}'R^{-1}\underline{1}) \quad (4.7.10)$$

and

$$\underline{1}'T^{-1}\underline{1} = \underline{1}'S^{-1}\underline{1}/(1 + t_s\underline{1}'S^{-1}\underline{1}) + \underline{1}'R^{-1}\underline{1}/(1 + t_r\underline{1}'R^{-1}\underline{1}). \quad (4.7.11)$$

[If R (and/or S) consists of the single element (h), $(h \geq 0$ and $t_r > 0)$, $R + t_r\underline{1}$ is the single element $t_r + h$. [The case $h = 0$ is required in 4.4 and $h > 0$ in 5.3.] The relevant terms in R in (4.7.8), (4.7.10) and (4.7.11) reduce simply to $(h + t_r)$, $\underline{x}^{(2)}/(h + t_r)$ and $1/(h + t_r)$

respectively, giving the initial terms in the iteration up the tree described in 4.4.]

5 · Further aspects of the problem and its likelihood solution

5.1 THE PROGRAM AND THE RESULTS

The program MAXTREE finds the maximum likelihood tree for a given set of data, using the principles laid down in the previous chapter. It has been extensively tested on hypothetical data and on two main sets of actual data. The first of these is a set of data on nine Asian and American populations (A and A), which is a subset of the data on 15 world-wide populations compiled by Cavalli-Sforza and Edwards and used extensively to test previous heuristic methods. Secondly there is a set of data on seven North-West-European populations (N. W. Eu.) provided by Professor J. H. Edwards. [Some of these data have been tabulated by Bjarnason et al. (1973).] Summaries of the data and diagrams and tables of the results obtained are given in figures and tables (b) and (c) of this section. A detailed description of the current form and performance of the MAXTREE program is provided, in order to demonstrate how the theoretical considerations of the previous chapter may be translated into practice, and to enable comparisons with other procedures to be made.

The program was first developed for the Cambridge Titan computer (an extensively modified Atlas I type computer) with an access time of 5 microseconds and 48 bit words (instructions). The program consists of 392 lines of FORTRAN (excluding COMMENT), which is less than half the length of Felsenstein's evaluation program and substantially shorter than the most recent (and shortest) version of Edwards' minimum evolution (ME) program (Edwards (1966), Thompson (1973a)). A flow diagram of the program is given in Fig. 5.1(a) and the approximate sizes of program and library routines, and running diagnostics, in Table 5.1(a). The program has also been modified for use on an IBM 370/165, but no extensive tests have been made. A factor of ten in the running time is to be expected, and this is the order of the factor obtained (Table 5.1(a)).

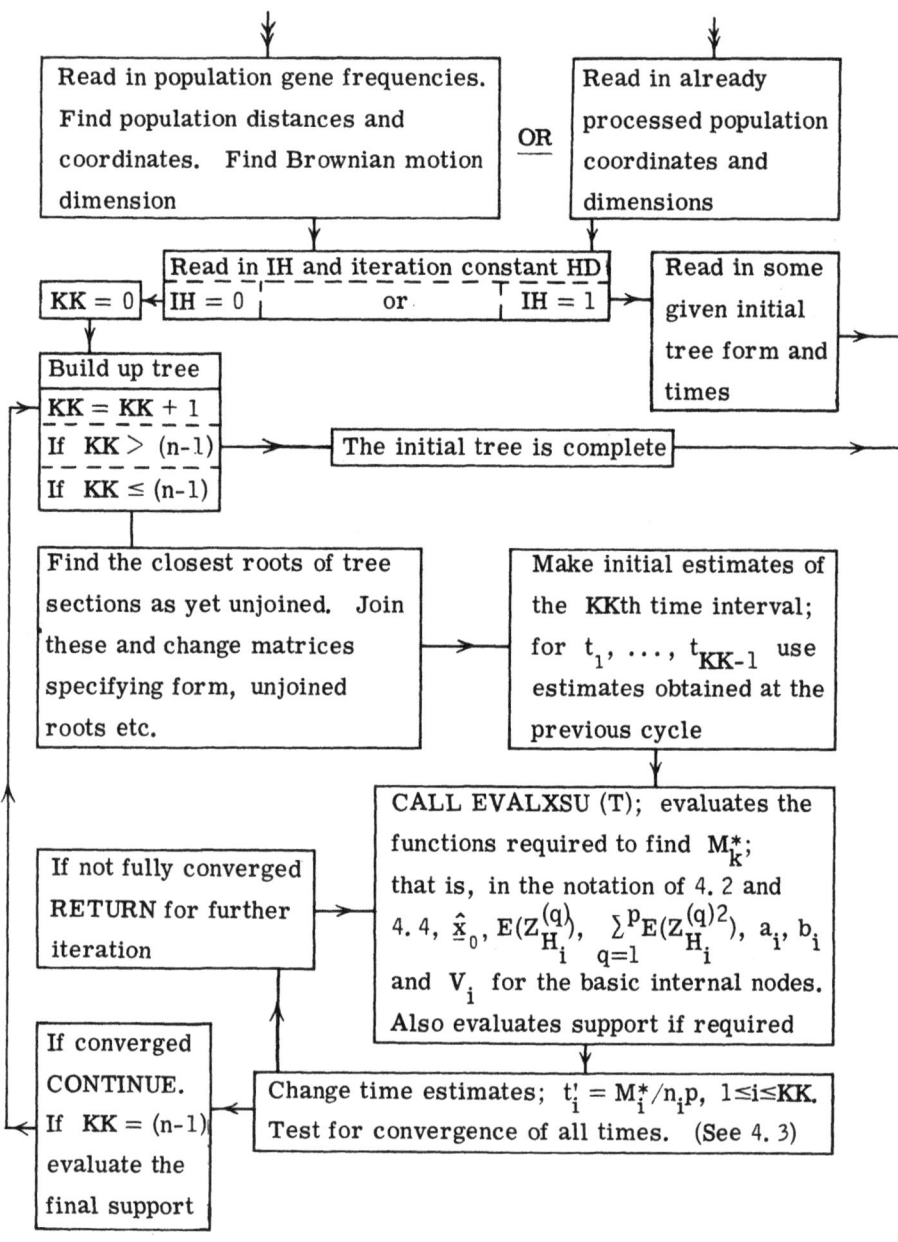

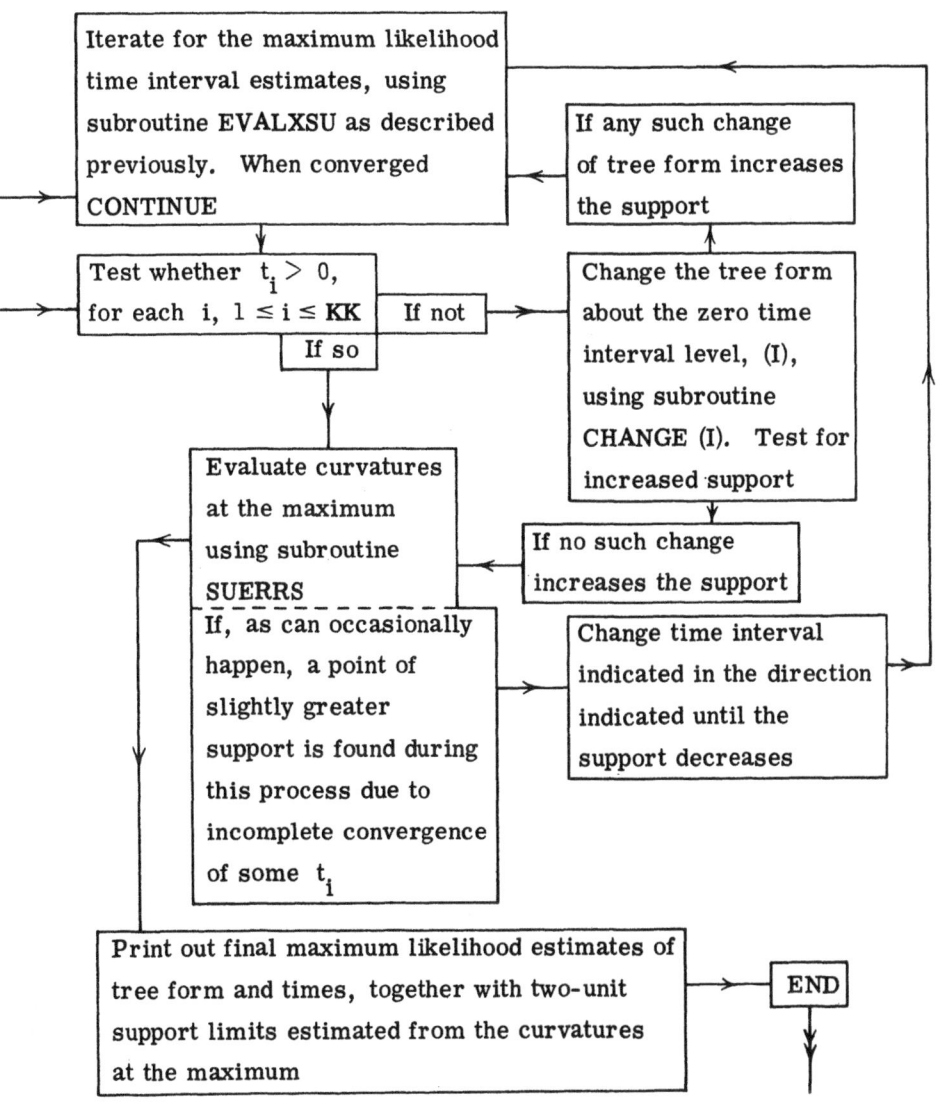

Fig. 5.1(a).   Flow diagram of the program MAXTREE.

The content of the boxes in the diagram:

Iterate for the maximum likelihood time interval estimates, using subroutine EVALXSU as described previously.   When converged CONTINUE

If any such change of tree form increases the support

Test whether $t_i > 0$, for each i, $1 \le i \le$ KK   If not

If so

Change the tree form about the zero time interval level, (I), using subroutine CHANGE (I).   Test for increased support

Evaluate curvatures at the maximum using subroutine SUERRS

If, as can occasionally happen, a point of slightly greater support is found during this process due to incomplete convergence of some t_i

If no such change increases the support

Change time interval indicated in the direction indicated until the support decreases

Print out final maximum likelihood estimates of tree form and times, together with two-unit support limits estimated from the curvatures at the maximum

END

Compile time	6. 87 secs
Run time	29. 47 secs
Approximate store required;	
MAIN program	1000 words
Subroutines	
EVALXSU	500 words
SUERRS	500 words
CHANGE	250 words
Titan library routines	3100 words
Common area (with array capacity for up to 12 populations)	829 words
TOTAL (with forward references, names etc.)	6702 words
	[322 K bits].
Note: on an IBM 370/165,	
Total store required	44 K bytes
	[352 K bits].
Run time (after compiling)	1. 5 to 2 secs

Table 5. 1(a).   Titan diagnostics of the MAXTREE program

The ME program mentioned above requires 14 K words of store and takes 25 seconds to compile and run for a data set of nine populations. Although this is much less in time and somewhat less in store than previous ME programs, the MAXTREE program requires the same order of time and only 7 K of store to accommodate the analysis of up to 12 populations.   Thus likelihood estimation is computationally more efficient than previous heuristic methods.   The run times for the A and A data (Table 5. 1(a)) are the largest so far obtained; some of the time intervals converge to zero (Table 5. 1(b)), giving cases of incomplete convergence (see below), and testing of alternative forms of tree.   A more usual time, for example for the seven N. W. Eu. populations, or for ten populations having a clearly defined tree form, is 15 to 20 seconds.

The program utilises population coordinates which must therefore be previously determined from the gene frequencies. This is more efficient than processing the gene frequencies within the program, since often several runs with the same data set will be required, using different subsets of the populations or different initial forms of tree. The gene frequency processing program (original version due to Edwards) allows for up to 15 populations and 10 gene loci with up to eight alleles at each. MAXTREE has no restriction on the Brownian motion dimension. There is an option either to build up a tree, or to start iteration from a given complete tree. This enables the stability of the maximum, and the likelihood of adjoining forms of tree, to be investigated.

If the tree is to be built up from the lowest level, the program at each stage joins the roots of distinct parts of the tree, starting at the populations and working up to the root. A few iterations are then performed to find the position of the root of the new subtree and approximate time interval estimates. This is not done to a high level of precision, since further iterations will be undergone at each level of building, and the addition of extra levels changes the ML time estimates. This procedure raises the question of stability of ML tree forms under addition of populations. Work done on the N.W. Eu. data reveals that although time estimates may change the tree form is reasonably stable, and together with opportunities for later change of form this method of finding an initial tree is certainly adequate.

The program then iterates for the ML tree. All the iteration constants (small numbers specifying required accuracy of roots) are multiples of a prescribed constant HD, which may be chosen fairly arbitrarily but should depend on the overall dispersion of the data, since it determines absolute, not relative, accuracy. When a change of tree form is indicated by the existence of a zero time interval estimate subroutine CHANGE effects this, tests whether or not the new form can give higher support and accepts or rejects it on these grounds (Fig. 5.1(a)). The procedure within CHANGE depends on whether or not the change is of topology as well as labelled history (4.5(ii)).

When a final tree form with an internal maximum is found, or when the current tree has zero intervals but no change about these produces a

The data used are a subset of the data on 15 world-wide populations compiled from various sources by A. W. F. Edwards and L. L. Cavalli-Sforza. These data have been used extensively to test previous heuristic methods of re-building evolutionary trees. There are data on five gene loci, $A_1 A_2 BO$, Rhesus, MNSs, Duffy (Fy) and Diego (Di) with 4, 7, 4, 2 and 2 alleles respectively, which together provide 14 Brownian motion dimensions.

Results: (n = 9, p = 14)

Maximum likelihood support; $\hat{S} = 159.6$

Limit of accuracy of ML estimates: $HD = 8 \times 10^{-5}$

Maximum likelihood time estimates, with their two-unit support limits estimated from the curvature at the maximum;

Time intervals $(\sigma^2 t_i)$	Time ago $(\sigma^2 s_i = \sigma^2 [\sum_{j=1}^{i} t_j])$
$i = 1;\quad (138 \pm 58) \times 10^{-4}$	138×10^{-4}
$i = 2;\quad (\ 22 \pm 61) \times 10^{-4}$	160×10^{-4}
$i = 3;\quad (\ 17 \pm 68) \times 10^{-4}$	177×10^{-4}
$i = 4;\quad (\ \ 2 \pm 44) \times 10^{-4}$	179×10^{-4}
$i = 5;\quad (148 \pm 93) \times 10^{-4}$	327×10^{-4}
$i = 6;\quad (\ 12 \pm 82) \times 10^{-4}$	339×10^{-4}
$i = 7;\quad (\ \ 0 \pm 34) \times 10^{-4}$	339×10^{-4}
$i = 8;\quad (\ \ 0 \pm 23) \times 10^{-4}$	339×10^{-4}

The two earliest time intervals have maximum likelihood estimate 0, giving a 4-way root. The populations thus fall into four groups.

Note: $\sigma^2 = 1/8N_e$, and hence if the total evolutionary time (s_{n-1}) is around 30,000 years, or 1,500 generations, N_e is of the order of 10^4.

Table 5.1(b). Results of the MAXTREE program for nine Asian and American populations.

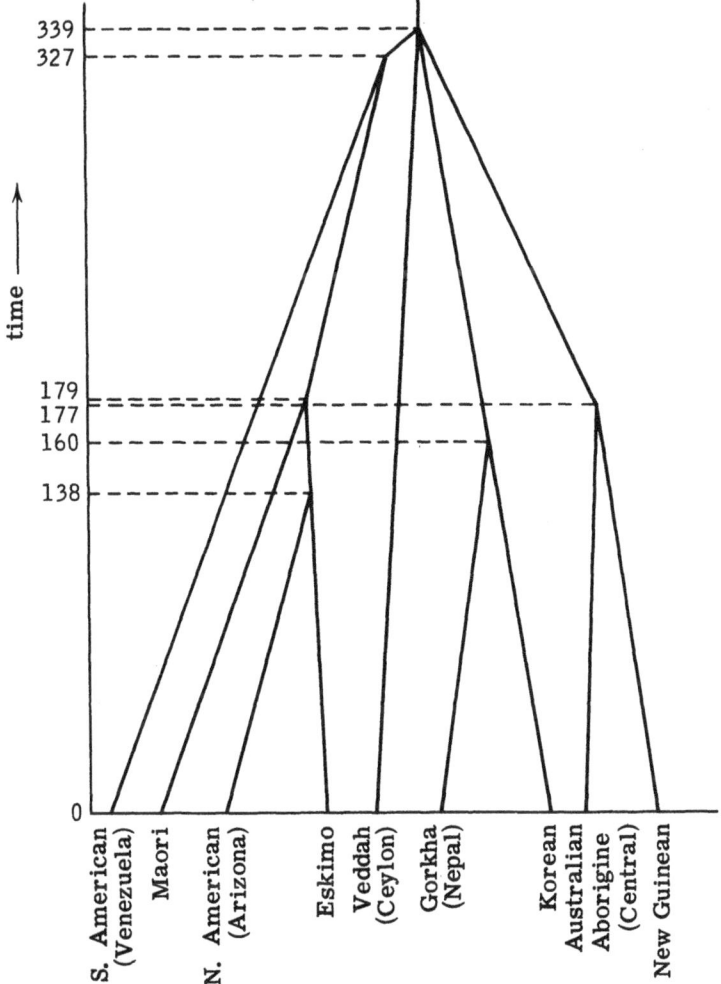

Fig. 5.1(b). Maximum likelihood tree for nine Asian and American populations. Diagram of maximum likelihood tree form and times of split. The ordering of populations is that given by the first principal component of position in the projected space. The scale of the time axis is in units of $(1/10^4 \sigma^2)$ generations, time being measured backwards from the present. Values of the support and maximum likelihood time interval estimates with their 2-unit support limits are given in Table 5.1(b), with a summary of the data used.

There are only five blood group loci for which data could be obtained for all sevel populations. These are ABO, Hp, Duffy (Fy), Kell and P, with 3, 2, 2, 2, and 2 alleles respectively. These together provide six Brownian motion dimensions. The data were provided by Professor J. H. Edwards. *

Results: (n = 7, p = 6)

Maximum likelihood support; $\hat{S}$ = 133.2

Limit of accuracy of ML estimates; HD = 2×10^{-5}

Maximum likelihood time estimates, with their two unit support limits estimated from the curvature at the maximum;

	Time intervals $(\sigma^2 t_i)$	Time ago $(\sigma^2 s_i = \sigma^2 [\sum_{j=1}^{i} t_j])$
i = 1;	$(5 \pm 16) \times 10^{-5}$	5×10^{-5}
i = 2;	$(24 \pm 16) \times 10^{-5}$	29×10^{-5}
i = 3;	$(8 \pm 29) \times 10^{-5}$	37×10^{-5}
i = 4;	$(8 \pm 38) \times 10^{-5}$	45×10^{-5}
i = 5;	$(54 \pm 67) \times 10^{-5}$	99×10^{-5}
i = 6;	$(37 \pm 82) \times 10^{-5}$	136×10^{-5}

The ML tree is bifurcating with no zero time interval estimates. The populations fall into two clearly distinguished groups; Norse and Celtic with England belonging to the former group and Iceland to the latter.

Note: $\sigma^2 = 1/8N_e$ and hence if the total evolutionary time of these populations (s_{n-1}) is around 4000 years or 200 generations, N_e is of the order of 2×10^4.

* Some of these data have now been tabulated by Bjarnason et al. (1973).

Table 5.1(c).   Results of the MAXTREE program for seven North-West-European populations.

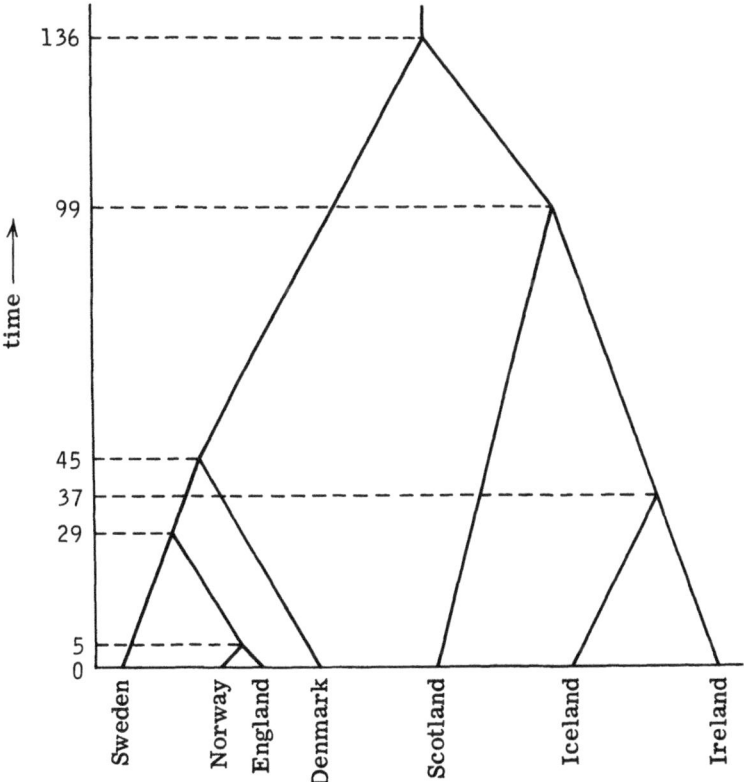

Fig. 5.1(c).   Maximum likelihood tree for seven North-West-
European populations. Diagram of the maximum
likelihood tree form and times of split. The ordering of populations is that given by the first principal
component of position in the projected space.   The
time axis is marked in units of $(1/10^5\sigma^2)$ generations, time being measured backwards from the
present. Values of support and maximum likelihood
time interval estimates with their 2-unit support
limits are given in Table 5.1(c), with a summary of
the data used.

form of greater support, control is transferred to subroutine SUERRS which estimates the curvature at the maximum by evaluating the support at points in the neighbourhood. Convergence of $\underline{t}$ is in terms of closeness to a root of $t_k = M_k^*/n_k p$ $(k = 1, \ldots, (n-1))$, no evaluations of support being necessary until the root is found. The factor $1/t_k^2$ in $\dfrac{\delta S}{\delta t_k}$ can mean that in cases of convergence to zero t_k can be approximately equal to $M_k^*/n_k p$ at some distance from the maximum. This occasionally results in insufficient convergence, and a point of higher support is found in the course of SUERRS. If this occurs the time interval in question is changed in the direction indicated until the support decreases and iteration is restarted at that point.

A complete matrix of coformations of the maximum relative support surface is not estimated, since this would require at least $\frac{1}{2} n(n + 1)$ evaluations. In fact the coformation between any two time intervals is negative at the maximum. In practice only curvatures in the directions of the axes are taken. The two values for each t_k (one taken in each direction) are combined to give a symmetric two-unit support limit. [If $\hat{t}_k < 10\,\text{HD}$ a single value in the increasing direction is used and this fact is indicated in the output.] In practice the two-unit support limits give little quantitative information. The support is not quadratic, nor even symmetric. However, besides a general indication that, these limits being large, the surface is very flat at the maximum, the limits do provide some measure of the relative degree of confidence in the different t_k, and also indicate whether, and which, other tree forms may be within two units of support of the maximum. For a large number of populations there may be many of these and only a few of the closest can be examined.

From the flow diagram (Fig. 5.1(a)) it may be seen that there are several theoretically non-ending loops. To prevent excessive looping only twenty iterations for each time interval, four alternative tree forms and five returns from SUERRS due to incomplete convergence are allowed. Only the first of these limits has ever been reached and this only in cases of convergence to zero. In this case the method of return from SUERRS is much quicker than allowing further iteration. The output from the program is of a form to be useful in further analysis rather than a final presentation of results. Each time a major stage is completed (a time iteration, form change or support evaluation) the relevant information is printed. This enables alternative forms, rates of convergence and non-symmetry of support limits to be seen, and these suggest possible tree forms to be tested in any further analysis.

Although the two-unit support limits, given in Tables 5.1(b), 5.1(c), for the two data sets include many other forms of tree, conver-gence to the ML form was obtained quite rapidly from a variety of differ-ent starting points. Further the forms obtained using subgroups of the populations were consistent with the overall form. This is not a necessary consequence of the model and to obtain such a result inspires confidence in the estimated forms. The comparison of trees produced by different data sets for the same populations is discussed below (5.2).

It is also possible to make estimates of N_e given independent anthropological estimates of s_{n-1} (or vice versa), and rough estimates are given in Tables 5.1(b), 5.1(c). The estimates are perhaps on the small side, but are certainly of the correct order of magnitude, indica-ting that the observed gene frequency differentiation could be the result of r.g.d. alone. It may be that small initial population sizes allowed rapid initial dispersion and dominate the overall effective population size. It is also likely that sampling has had some effect in increasing the appar-ent dispersion.

5.2 THE BIG-BANG LIKELIHOOD

We have computed maximum likelihood estimates of evolutionary trees, and maximum support values. However these do not give an ade-quate picture of the shape of the likelihood surface, the support for alter-natives, or the amount of information on phylogenetic relationships con-tained in a set of data. The curvatures at the maximum give a measure of confidence in the time interval estimates, but in practice these are estimated only in the direction of the axes and convey little except that the confidence limits are wide.

We define the Big-Bang (BB) tree as the tree in which all popula-tions were formed simultaneously by the splitting of a single ancestor and have since evolved independently: $t_i = 0$ for all $i \neq 1$ and F is unspecified since all F give the same tree when all time intervals between splits are zero.

Then

$$\hat{x}_0^{(q)}(\underline{t}) = [\sum_{i=1}^{n} x_i^{(q)}]/n = \bar{x}^{(q)} \tag{5.2.1}$$

and $-2S(\hat{\underline{x}}_0(\underline{t}), \sigma^2\underline{t}, F) = np\log(\sigma^2 t_1) + (1/\sigma^2 t_1)\sum_{i=1}^{n} \|\underline{x}_i - \bar{\underline{x}}\|^2.$

Thus

$$\widehat{\sigma^2 t}_1 = (1/np)\sum_{i=1}^{n} \|\underline{x}_i - \bar{\underline{x}}\|^2 = X^2/np, \tag{5.2.2}$$

where X^2 is the total dispersion of the populations.

Then

$$S(BB) = S(\hat{\underline{x}}_0, (\widehat{\sigma^2 t}_1, 0, \ldots, 0), F) = -\tfrac{1}{2}np(1 + \log(X^2/np)). \tag{5.2.3}$$

It is not suggested that BB is a likely hypothesis, but it provides a useful basic reference point for a given set of data, since were it the ML solution no inferences about F could be made. Only support differences between different hypotheses, on the same data, have meaning; support differences between maximising hypotheses for different data sets do not.

The difference in support

$$\delta S = S(ML) - S(BB) = \hat{S} - S(BB) \tag{5.2.4}$$

provides a measure of the amount of information in the data, and hence enables comparisons between different data sets to be made.

We may then go further and compute the support for trees having r non-zero splitting intervals, $r = 1, 2, \ldots$. In this way the 'simplest' hypothesis compatible with the data may be found, where this hypothesis is the H with smallest possible r satisfying

$$\hat{S} - S(H) \leq 2$$

(or some other predetermined level). [I am indebted to Professor D. R. Cox for this suggestion.] However, even for $r = 1$ there are too many alternative F to consider. Thus instead we start at the ML hypothesis, and then consider suitable restrictions $t_i = 0$. It is found that the support is sensitive to changes in the total evolutionary time, T $(= \sigma^2 s_{n-1})$, but is far less so to changes in t_i subject to constant T, particularly for $i = (n-1)$, $(n-2)$. In the case of the N. W. Eu. data the support scarcely alters along the line; $\sigma^2 t_i = \widehat{\sigma^2 t_i}$ $i = 1, \ldots, 4$ and $\sigma^2(t_5 + t_6) = 90 \times 10^{-5}$, from $\sigma^2 t_6 = 0$ to $\sigma^2 t_6 = 50 \times 10^{-5}$.

Some further results are given in Tables 5.2(a), 5.2(b). We see

$n = 9$, $p = 14$, $X^2 = 4.1677$

Hypothesis H	Support difference $\hat{S} - S(H)$	Evolutionary time; $\hat{T}$
Maximum likelihood $(\hat{t}_7 = \hat{t}_8 = 0)$	0	0.0339
Big Bang $(t_i = 0, i \neq 1)$	8.0	0.0331
Maximum for the ML form F subject to $t_i = 0$ for:		
$i = 4, 6, 7, 8$	0.1	0.0338
$i = 3, 4, 6, 7, 8$	0.2	0.0337
$i = 2, 3, 4, 6, 7, 8$	0.4	0.0338

Likelihood ratio for ML tree over Big Bang tree $= 10^{3.5}$

Trees with $t_i = 0$ for $i \neq 1$ and 5 have support values close to the maximum. If $t_i = 0$ for $i = 2, 3, 4, 6, 7$ and 8;

$$\sigma^2 \hat{t}_1 = 0.0168, \quad \sigma^2 \hat{t}_5 = 0.0170.$$

Table 5.2(a). Supports for hypotheses of simultaneous splitting; American and Asian data. See Fig. 5.1(b) and Table 5.1(b).

that many hypotheses of simultaneous splitting do not have significantly lower support than the maximum, although for neither set of data is BB a tenable hypothesis. The estimate $\hat{T}$ is scarcely changed by restrictions $t_i = 0$. Although the support differences are small the iterative method gives good discrimination of which tree form should be chosen and which time intervals made non-zero.

The N. W. Eu. populations give a much larger value to δS, (5.2.4), than do the A and A populations, showing that the former contain more information on phylogenetic relationships. That this is so is shown also by the fact that for the former data the ML tree is bifurcating while for the

$n = 7$, $p = 6$, $X^2 = 0.0551$

Hypothesis H	Support difference $\hat{S} - S(H)$	Evolutionary time; $\hat{T}$
Maximum likelihood $(\hat{t}_i > 0,$ i)	0	0.00136
Big Bang $(t_i = 0,$ i $\neq$ 1)	14.8	0.00131
Maximum for the ML form (F) subject to $t_i = 0$ for;		
i = 6	0.05	0.00132
i = 3, 4	0.2	0.00134
i = 3, 4, 6	0.2	0.00133
i = 5, 6	2.2	0.00133
i = 2, 3, 4	3.5	0.00135
i = 2, 3, 4, 6	3.7	0.00132

Likelihood ratio for ML tree over the Big Bang tree = $10^{6.5}$

Trees with $t_i = 0$ for i = 3, 4 and 6 have support values close to the maximum. If $t_3 = t_4 = t_6 = 0$;

$$\sigma^2 \hat{t}_1 = 0.00005, \quad \sigma^2 \hat{t}_2 = 0.00032, \quad \text{and} \quad \sigma^2 \hat{t}_5 = 0.00096.$$

Table 5.2(b). Supports for hypotheses of simultaneous splitting; European data. See Fig. 5.1(c) and Table 5.1(c).

latter the ML solution has a four-way root. Thus the A and A data provide no information on the relationships between the four major subtrees. Note that, as for all measures of statistical information, there is no assumption that the information is correct. Only validity of the model and large accurate data sets can ensure the reliability of inferred phylogenetic relationships.

From Table 5.2(a) the most that can be inferred with any degree of statistical confidence is $t_1 > 0$, $t_5 > 0$, giving a tree similar to the ML tree in form with a 5-way root. Around mid-way in time between the root and the present three of these five populations split giving rise to the three more closely related groups; Gorkha and Korean; Australian and New Guinean; North American, Maori and Eskimo. For the N.W. Eu.

populations the split of Iceland and Ireland from the Norse populations can
be reliably inferred although the position of Scotland is less clear (Table
5. 2(b)). Whether Ireland and Iceland or the Norse populations are the
more closely related cannot be reliably inferred, but for this particular
rather limited set of data it seems that England is significantly more
closely related to Norway than to any of the other populations.

Although within any given problem the degrees of freedom are not
relevant, since we are interested only in which point hypotheses do fall
within two units of support of the maximum (1. 3), in a quantitative comparison of different data sets the effect of varying p and n should be
considered. For the BB tree (n - 2) constraints are placed on the t_i,
and, were classical asymptotic likelihood ratio testing theory applicable,
it would assign a $\frac{1}{2}\chi^2_{n-2}$ significance test to δS. On the alternative
hypothesis the non-centrality of the χ^2 is the distance of the true hypothesis from BB in the metric of the information matrix. Given that BB
is not a tenable hypothesis, it is the difference between the distances from
BB provided by the different data sets that we wish to consider.

The classical approach as such does not apply, since as p increases so also does the number of parameters and asymptotic consistency and Normality will not obtain, but we may consider the above
'distances' from BB. An additional gene locus clearly gives additional
information in some sense, and will increase δS if this confirms the
phylogenetic relationships previously inferred; asymptotically, for fixed
n, the increase should be linear in p. Both the general support function
and S(BB), (5. 2. 3), are of the order np for given $\sigma^2 s_{n-1}$, and computations show that, at least for small n and p, δS increases with both n
and p. Thus $\delta S/np$ the information/population/dimension appears to be
the appropriate measure of the relative information in different sets of
loci with regard to the phylogenetic relationships between different sets
of populations.

There are insufficient data to examine extensively the agreement
of trees produced using different sets of gene loci for the same populations.
Individual loci do not give good agreement, but this is not to be expected
(2. 4). Data for five of the European populations (Norway and Sweden ex-
cluded) is available for five further gene loci, also giving p = 6. The ML

tree for these data confirms the major Celtic-Norse split but not the other details of the tree. Substantial agreement is not to be expected for this small value of p, and the importance of differences of detail is difficult to assess. However it is to be hoped that, were such data available, large accurate data sets for loci and populations for which random genetic drift is accepted as the major differentiating force would give the same estimate of the tree form.

	Data 1	Data 2
Loci (s = 5)	ABO, Hp, Fy, Kell, P.	AcPh, Gm, MN, Luth, PGM.
Number of alleles (k_i)	3, 2, 2, 2, 2	3, 2, 2, 2, 2
$p = \sum_{i=1}^{s} (k_i - 1)$	6	6
Dispersion $= X^2$	0.0412	0.0325
ML support $= \hat{S}$	88.6	88.2
BB support $= S(BB)$	83.9	87.8
$\delta S = \hat{S} - S(BB)$	4.7	0.4

Table 5.2(c).   The comparison of two data sets giving data for the same five populations: England, Denmark, Scotland, Ireland and Iceland.

The support difference δS provides a measure of the relative amounts of information in the two data sets, which have the same n and p (Table 5.2(c)). The first set of loci contains significant information: the second does not. Although the two maximum supports differ only slightly it is δS that is the relevant factor. It is only the introduction of some basic reference point that enables comparisons to be made, and we suggest BB as a useful and meaningful such reference point.

5.3 DISTORTIONS OF THE TIME SCALE

Although it seems that valid inferences as to tree form, F, may often be made, the time estimates are less reliable. The two main causes of distortion of the time estimates are variation in population size and sampling. There is also the constant scaling factor caused by stereographic projection (2.3.8), but this is negligible, and for any given group

of populations may be corrected for. We show first that constancy of
population size, in time, is not necessary for the validity of the model.

Suppose that the size at time s ago, of each population then
existent is $N_e(s)$, and let

$$u_i = \int_0^{s_i} [ds/8N_e(s)], \quad i = 1, \ldots, (n-1). \tag{5.3.1}$$

Consider the diffusion $\underline{z}$ on the arc between type 0 nodes at times s_i
and s_{i*} ago.

That is $\underline{z} \doteq \underline{y}_j - \underline{y}_j^*$ for some type 0 node (or population) $\underline{y}_j$, and
its immediate ancestor $\underline{y}_j^*$ (which may be the root $\underline{x}_0$).

Then $z^{(q)}$ is $N(0, \int_{s_i}^{s_{i*}} [ds/8N_e(s)])$ or $N(0, u_{i*} - u_i)$ $q = 1, \ldots, p,$
and all such diffusions are independent.

$\underline{x}_i - \underline{x}_0$ is the sum of such independent diffusions $(i = 1, \ldots, n)$;
hence $\underline{x}^{(q)}$ is multi-variate Normal and

$$\text{cov}(x_i^{(q)}, x_j^{(q)}) = u_{n-1} - u_{l(i, j)} \quad \text{(cf. (3.1.1)).}$$

Thus the likelihood $L(\underline{x}_0, \underline{u}, F)$ is precisely as before, (3.1.2), with
$\underline{u}$ replacing $\sigma^2 s$. We may refer to u as <u>effective evolutionary time.</u>

Thus inferences as to tree form are unaffected by changes in N_e.
No absolute time scale may be inferred, but estimates of absolute effective
evolutionary times may be made. Estimates of actual time may only be
made if $N_e(s)$ is known: we may have independent evidence as to its
order of magnitude. The time scale at any point in history is in units of
$8N_e(s)$ generations, irrespective of whether this varies with s, and this
may be taken into account in comparing time intervals at different stages
in history.

Differences between population sizes at a single point in time will
cause distortions that are less readily eliminated. [I am indebted to
Professor C. A. B. Smith for causing me to consider this question.]
The above considerations show that the effect of differing population sizes
on the support function is that of a 'weight' proportional to the effective
population size over the given time interval attached to each arc of the
tree. If the population sizes at all stages in history were known, it would

in theory be possible to introduce a weighting procedure into the iterative method for estimating a tree. These weighting factors would affect estimates of both the means and variances of internal node positions, and the definition of M_k must be modified to incorporate them, but the fundamental result (4.3.5) and the method of Chapter 4 would then still apply.

Such a procedure may not however be computationally feasible, and is in any case not to be recommended. The effect of incorporating differential sizes on the form of the support function and on the properties of the iterative method are not known, and the effect of errors in relative population size cannot be estimated. The current relative effective population sizes will rarely be accurately known, and those pertaining in history probably never. We must therefore retain the restriction to equal population sizes at any given time: nevertheless the above considerations allow us to recognise that the analysis attaches undue weight to the position of small populations, and in any given situation the qualitative effect on the estimated tree may be considered.

The genetic aspects of sampling have been previously discussed (2.2 and 2.4); we consider here its effect on likelihood inferences. So far the population frequencies have been assumed known; in practice they are estimated from samples, size $m_i^{(h)}$ from population i at locus h say. At best, when all genotypes are distinguishable, this is a multinomial sampling (assuming sampling 'with replacement' or $m_i^{(h)} \ll N_e$). More often we have only maximum likelihood estimates of gene frequency from phenotype data. However, provided there are more phenotypes than alleles the situation remains approximately multinomial.

In this case we have a sample of $2m_i^{(h)}$ genes, and the estimated frequencies $\hat{\underline{p}}$ are $(1/2m_i^{(h)})M(2m_i^{(h)}; \underline{p})$ where $\underline{p}$ is the true population frequency and M denotes a multinomial variate of the given index and parameter. Performing the transformations of 2.3 we obtain, by analogy with the multinomial sampling of r.g.d., for large $m_i^{(h)}$ (but $m_i^{(h)} \ll N_e$) $\hat{x}_i^{(q)}$ is $N(x_i^{(q)}, 1/8m_i^{(h)})$, where $\hat{\underline{x}}_i$ are the observed, and $\underline{x}_i$ the true, population positions, and q is a dimension corresponding to locus h. Further all $\hat{x}_i^{(q)}$ are independent, given the true population positions $\underline{\underline{x}}$.

Then $\hat{\underline{x}}^{(q)}$ is $N(x_0^{(q)}\underline{1}, \sigma^2 T + \text{diag.} (1/8m_i^{(h)}, 1 \le i \le n))$

$$= N(x_0^{(q)}\underline{1},\ \sigma^2(T + \text{diag. } (N_e/m_i^{(h)},\ 1 \le i \le n))),$$

where $T_{ij} = \sum\limits_{r=(i,\,j)+1}^{n-1} t_r,$ $\sigma^2 = 1/8N_e,$ and diag. ( ) denotes a diagonal matrix with the given components. $\hat{\underline{x}}^{(q)}$ $(q = 1, \ldots, p)$ are independent. σ^2, or N_e, remains a scaling factor for $\underline{t}$ but not for the total variance. The variance scaling factor (2.3.8) is the same for both the sampling and the r.g.d., and again may be corrected for at each locus, for populations in a given region of space.

Lemma. With the notation as defined in the previous chapter, the maximum likelihood estimates $\hat{t}_i$ $(i = 1, \ldots, (n-1))$ satisfy

$$\hat{t}_i = E(\ _i|\hat{\underline{\underline{x}}},\ \hat{\underline{x}}_0(\hat{t}),\ \hat{\underline{t}},\ F)/n_i p, \tag{5.3.2}$$

where $\hat{\underline{\underline{x}}}$ is the set of observed population positions, and $\hat{\underline{x}}_0(t)$ is the maximum likelihood estimate of $\underline{x}_0$ for given $\underline{t}$ (and F). (The support remains quadratic in each $x_0^{(q)}$ and thus $\hat{\underline{x}}_0(\underline{t})$ may be found explicitly as in the previous chapter.)

Proof. Given $\underline{x}$, $\hat{\underline{\underline{x}}}$ is independent of $\underline{t}$, $\underline{x}_0$ and F, and $\underline{y}$ and $\hat{\underline{\underline{x}}}$ are independent. Then

$$L(\underline{x}_0,\ \underline{t},\ F) = f(\hat{\underline{\underline{x}}}|\underline{x}_0,\ \underline{t},\ F)$$

$$= \int \ldots \int_{\underline{x}} f(\hat{\underline{\underline{x}}}|\underline{x})\, f(\underline{x}|\underline{x}_0,\ \underline{t},\ F)\, d\underline{x}$$

$$\frac{\delta L}{\delta t_k} = \int \ldots \int_{\underline{x}} f(\hat{\underline{\underline{x}}}|\underline{x})\, \frac{\delta f(\underline{x}|\underline{x}_0,\ \underline{t},\ F)}{\delta t_k}\, d\underline{x}$$

$$= \int \ldots \int_{\underline{x}} f(\hat{\underline{\underline{x}}}|\underline{x})[-\tfrac{1}{2} f(\underline{x}|\underline{x}_0,\underline{t},F)[n_k p/t_k) - E(C_k|\underline{x},\underline{x}_0,\underline{t},F)/t_k^2]]d\underline{\underline{x}}$$

(from 4.3)

$$= -\tfrac{1}{2}[n_k p L/t_k - \int \ldots \int_{\underline{x},\,\underline{y}} C_k\, f(\hat{\underline{\underline{x}}},\underline{x},\underline{y}|\underline{x}_0,\underline{t},F)\, d\underline{x}\, d\underline{y}/t_k^2]$$

$$= -\tfrac{1}{2}[n_k p L/t_k - L \int \ldots \int_{\underline{x},\,\underline{y}} C_k\, f(\underline{x},\underline{y}|\hat{\underline{\underline{x}}},\underline{x}_0,\underline{t},F)\, d\underline{x}\, d\underline{y}/t_k^2]$$

$$= -\tfrac{1}{2}(n_k p/t_k^2)L(\underline{x}_0,\ \underline{t},\ F)[t_k - E(C_k|\hat{\underline{\underline{x}}},\ \underline{x}_0,\ \underline{t},\ F)/n_k p]$$

and hence for a stationary point w.r.t. $\underline{t}$ and $\underline{x}_0$

$$t_k = E(C_k|\hat{\underline{\underline{x}}},\ \hat{\underline{x}}_0(\underline{t}),\ \underline{t},\ F)/n_k p, \quad \text{(cf. 4.3).} \ /\!/$$

But

$$E(C_k | \hat{\underline{x}}, \underline{x}_0, \underline{t}, F) = E(E(C_k | \underline{x}, \underline{x}_0, \underline{t}, F) | \hat{\underline{x}}, \underline{x}_0, \underline{t}, F)$$

$$= E(M_k | \hat{\underline{x}}, \underline{x}_0, \underline{t}, F) \quad \text{(by definition)} \quad (5.3.3)$$

M_k is a quadratic function of $\underline{x}$, and (5.5.3) can in theory by found from the conditional distribution of $\underline{x}$, given $\hat{\underline{x}}$ and the parameters. If sample sizes are equal for all loci, the iterative method of Chapter 4 may be readily modified to give the true ML solution, for in this case the variance does not differ between dimensions and any orthogonal transformation of the projected coordinates is equivalent to the original data (4.6(i)). In iterating up the tree using (4.7.10) and (4.7.11), as described in 4.4 we must simply change the 'covariance matrix' for any 'subtree' consisting of a single population node from the previous single element (0) to the element $(1/8m_i)$ $(m_i = m_i^{(h)}$ for all h. See 4.4 and Appendix 2 of 4.7.) A restriction $t_1 \geq 0$ must also be included but has no effect on the iterative method; the point $t_1 = 0$ is no longer a singularity since sampling provides a positive lower bound to the population variances. Convergence properties, etc., will be similar to those described in Chapter 4, but the program of 5.1 has not been modified in this way, since sample sizes, even where stated, are not equal for all loci, except perhaps where a single recent study of a population has covered many blood group systems. If sample sizes are unequal the theory remains the same, but it would be necessary to consider the basic projected coordinates and not a transformation based only on distances; this is not at present computationally feasible.

The above modification of the method corresponds to Felsenstein's observation (personal communication) that sampling, with sample size m_i, is equivalent to N_e/m_i generations (or $1/8m_i$ units of $1/\sigma^2$ generations) of evolution. He notes that the ML solution with sampling included is the same as the solution without sampling for the same population positions at times $1/8m_i$ in the future. His proposal, equivalent to the above, is that the ML solution for the populations at these unequal hypothetical time points be found, and then the extra times subtracted off to reobtain the contemporary populations. The required restriction that the time of the last ancestral split occurs no later than the present prevents the singularity which is in general caused by populations at unequal time points (cf. 4.6(ii)). Although, for our iterative method, the modification to the single element matrices described above provides the simplest method of solving the problem, Felsenstein's representation of the situation provides a clearer idea of the effect of the inclusion of sampling on the estimated tree and on the relative support for alternative estimates. The smaller the sample size the less is our knowledge of the true present population position; equivalently the further into the future is the 'effective time' of the sample point.

The situation in which all m_i are equal (say to m) provides the particularly simple case in which the 'effective times' of the populations remain equal. The only effect of the inclusion of sampling is to reduce the estimate of $\sigma^2 t$ by an amount $1/8m$ (provided previously $\widehat{\sigma^2 t} > 1/8m$), for the term $\sigma^2 t_1 + 1/8m$ replaces $\sigma^2 t_1$ in the diagonal terms of the matrix $\sigma^2 T$, and $\sigma^2 t_1$ does not appear elsewhere in the support function. Thus when all sample sizes are of the same order of magnitude sampling should have little effect on the estimated form of tree.

However, in practice the variation between loci is often much greater than between populations and an alternative suggestion is the following. The coordinates used are those obtained by embedding the genetic distances in any Euclidean space (4.6(i)). But the observed squared distance $d_{ij}^{(h)2}$ at locus h, between populations i and j, with sample sizes $m_i^{(h)}$, $m_j^{(h)}$ satisfies

$$E(d_{ij}^{(h)2}) = (k_h - 1)((1/8m_i^{(h)}) + (1/8m_j^{(h)})) + d_{0,ij}^{(h)2} \qquad (5.3.4)$$

for $h = 1, \ldots, s$, where $d_{0,ij}^{(h)2}$ is the (unknown) true squared distance between populations i and j at locus h, having k_h alleles. Thus we may define

$$d_{ij}^{*2} = d_{ij}^2 - \sum_{h=1}^{s} (k_h - 1)(1/8m_i^{(h)} + 1/8m_j^{(h)}), \qquad (5.3.5)$$

where d_{ij}^2 is the total observed distance between populations i and j, and embed the modified distances d_{ij}^{*2} $(i, j = 1, \ldots, n)$ in the Euclidean space. This may be the best method in practice where sample sizes differ widely with both loci and populations. The tree inferred is, at least, the correct likelihood solution for some set of true population distances, the squared distances being unbiassed estimates of the true squared distances based on the observed data.

However, unless samples are large, we may obtain distances which do not satisfy the Euclidean metric conditions, we may even have negative distances between populations where there is no evidence of differing gene frequency. Thus although the method is useful in practice there may be problems. In fact, although estimated distances at some loci may be negative, the total distances (5.3.5) are usually positive and can often be embedded in a Euclidean space.

5.4 THE MISSING DATA PROBLEM

Although some heuristic methods are based entirely on distances and can use any available data to compute these, any method based strictly on the probability model must either use loci for which all data are present, or else take account of missing coordinates in some logically justifiable way. Having inferred an evolutionary tree we may obtain the probability distribution of missing coordinates. We then have the following problem: 'To what extent are the maximum probability estimates (or means) of unobserved random variables compatible with the ML values of the parameters on which they are based?'

Suppose we have a subset $\underline{x}_{\underline{g}}$ of $\{x_i^{(q)}; i = 1, \ldots, n,$
$q = 1, \ldots, p\}$ with only r_q of the coordinates in dimension q present
$(1 \leq r_q \leq n)$. Suppose further that we with to infer $(\underline{x}_0, \underline{t}, F)$ using all available data. Population distances are no longer sufficient, and we must consider the projected coordinates. For each q $(1 \leq q \leq p)$ $x^{(q)}\big|_{\underline{x}_{\underline{g}}}$ is

$$N(x_0^{(q)}\underline{1}, T_q^*), \tag{5.4.1}$$

where $\underline{x}^{(q)}\big|_{\underline{x}_{\underline{g}}}$ denotes the restriction of $\underline{x}^{(q)}$ to $\underline{x}_{\underline{g}}$ and T_q^* is the
r_q by r_q matrix formed by eliminating from T those rows and columns corresponding to populations for which there are no data in dimension q. Then

$$\hat{x}_0^{(q)}(\underline{x}_{\underline{g}}, \underline{t}, F) = ([\underline{x}^{(q)}\big|_{\underline{x}_{\underline{g}}}]' T_q^{*-1}\underline{1})/(\underline{1}'T_q^{*-1}\underline{1}). \tag{5.4.2}$$

114

Estimation of $\underline{t}$ and F remains theoretically possible, although it is complicated by the fact that some dimensions no longer contain information on each t_j separately. However, as may be shown by a proof similar to that of (5. 3. 2), we still have the fundamental result that

$$\hat{t}_{k^n_k}p = E(C_k | \underline{x}_g, \hat{x}_0(\underline{x}_g, \hat{t}, F), \hat{t}, F), \quad k=1, \ldots, (n-1). \quad (5.4.3)$$

Having obtained estimates of $(\underline{x}_0, \underline{t}, F)$ based upon $\underline{x}_g$ we may consider the missing coordinates $\underline{x}_m$. Since the dimensions are independent we consider each separately and drop the superscript (q). We call the ML tree for those populations for which there are data in the given dimension the 'framework' of the tree (Fig. 5. 4). Let $w = n - r_q$.

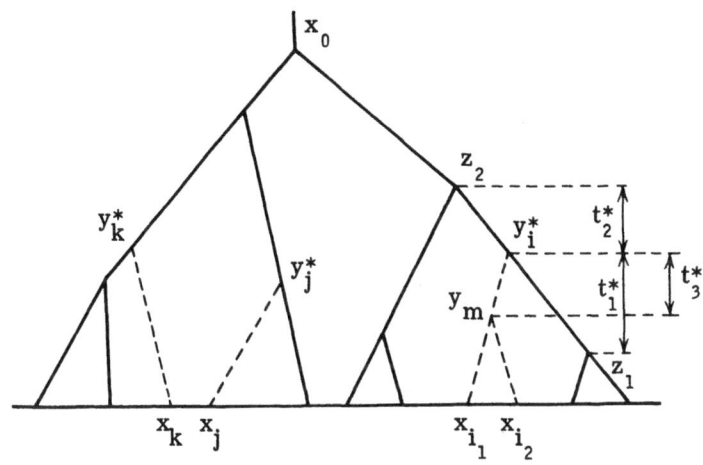

Fig. 5. 4. The estimation of missing data coordinates. The framework of the tree is shown by the solid lines, while the broken lines denote missing parts of the tree.

Given $\underline{x}_g$ and the parameters, $\underline{x}_m$ is w-variate Normal. Without loss of generality $\underline{x}_m = (x_1, \ldots, x_w)$ and $\underline{x}_g = (x_{w+1}, \ldots, x_n)$. Let y_i^* be the last ancestor of x_i on the framework of the tree $(i=1, \ldots, w)$; Fig. 5. 4.

Then the maximum probability estimate (or mean) is

$$\hat{x}_i = E(x_i | \underline{x}_{-g}, \ x_0, \ \underline{t}, \ F)$$

$$= E(E(x_i | \underline{x}_{-g}, \ x_0, \ \underline{t}, \ F, \ y_i^*) | \underline{x}_{-g}, \ x_0, \ \underline{t}, \ F)$$

$$= E(y_i^* | \underline{x}_{-g}, \ x_0, \ \underline{t}, \ F). \tag{5.4.4}$$

Lemma. If $x_i = E(y_i^* | \underline{x}_{-g}, \ x_0, \ \underline{t}, \ F)$ $(i = 1, \ldots, w)$ then

$$E(y_i^* | \underline{x}_{-m}, \ \underline{x}_{-g}, \ x_0, \ \underline{t}, \ F) = x_i, \tag{5.4.5}$$

and

$$\hat{x}_0(\underline{x}_{-g}, \ \underline{t}, \ F) = \hat{x}_0(\underline{x}_{-m}, \ \underline{x}_{-g}, \ \underline{t}, \ F). \tag{5.4.6}$$

Proof. Since all mean positions and $\hat{x}_0$ may be determined recursively by the series of equations relating each node to its immediate neighbours (4.4), it is sufficient to consider only the equations relating y_i^* to its immediate neighbours. The equations relating framework splitting points to their neighbours are necessarily unchanged.

Now given z_1 and z_2 of Fig. 5.4 and time intervals as shown,

$$\hat{x}_{i_1} = \hat{x}_{i_2} = \hat{y}_m = E(y_i^* | z_1, \ z_2, \ \underline{t}, \ F) = (t_2^* z_1 + t_1^* z_2)/(t_1^* + t_2^*)$$

But

$$E(y_i^* | y_m, z_1, z_2, \underline{t}, F) = ((1/t_1^*) + (1/t_2^*) + (1/t_3^*))^{-1}((z_2/t_2^*) + (z_1/t_1^*) + (y_m/t_3^*))$$

$$= \hat{y}_m = \hat{x}_{i_1} = \hat{x}_{i_2} \ \text{if} \ y_m = \hat{y}_m$$

$$= E(y_i^* | z_1, \ z_2, \ \underline{t}, \ F). \ /\!/$$

Thus we have that

(i)   Missing coordinates should be estimated by the mean positions of their ancestors on the framework of the tree,

(ii)   Given $\underline{t}$ and F, estimates of all node positions (including x_0) are unchanged by the introduction of the missing coordinates at their mean positions, and introducing them in this way is 'consistent' with node positions already inferred.

However, the ML estimates of the parameters $(\hat{t}, \ \hat{F}$ and $\hat{x}_0$ $(= \hat{x}_0(\underline{x}, \ \hat{t}, \ \hat{F})))$, are not unchanged by this insertion of the missing data; the ML solution does not usually have 'vertical' arcs in any part of the

116

tree. The question of change of ML tree form under the inclusion of additional populations, discussed in 5.1 with reference to the adequacy of the method of constructing an initial tree, is closely related to the problem here. The true ML tree inferred from the data is not precisely the same as the 'revised' tree inferred from the data and additional hypothetical data constructed on the basis of that ML tree: but such data do not usually cause extensive changes in the tree inferred.

The approach here may be compared with Fisher's 'predictive likelihood' (Fisher (1956, p. 126)), which would suggest that the complete probability distribution

$$f(\underline{x}_m, \underline{x}_g | \underline{x}_0, \underline{t}, F) \qquad (5.4.7)$$

should be regarded as a likelihood for $(\underline{x}_m, \underline{x}_0, \underline{t}, F)$ and maximised jointly with respect to these variables. This approach leads to singularities if applied to the set of all internal nodes, but in this case is feasible. Because of Normality, the tree inferred by maximising (5.4.7) is an equilibrium limit of a series of repeatedly 'revised' trees.

5.5 ANCILLARITY AND THE NUISANCE PARAMETER $\underline{x}_0$

As yet we have considered only the complete likelihood $L(\underline{x}_0, \underline{t}, F)$, and the MRL

$$L^*(\underline{t}, F) = \max_{\underline{x}_0} . [L(\underline{x}_0, \underline{t}, F)]. \qquad (5.5.1)$$

We have discussed the disadvantages of eliminating any parameter other than by considering the MRL (1.3), and have also seen that $\underline{x}_0$ should sometimes be considered, for example in the missing data problem or in any situation where the node positions are of interest. However, in some situations it may be that we are interested only in inferring the evolutionary history $(\underline{t}, F)$, and $\underline{x}_0$ may be truly a nuisance parameter, although without $\underline{x}_0$ the estimated tree cannot be fully specified. The method of Felsenstein (1973) suggests a possible procedure for the elimination of $\underline{x}_0$. [The notation and terminology of Felsenstein (1968, 1973) are here slightly modified to correspond with those of Chapters 3 and 4.]

Felsenstein (1968) shows how for each q $(1 \leq q \leq p)$ and for given $(\underline{t}, F)$ statistics $u_i^{(q)} = u_i(\underline{x}^{(q)}; \underline{t})$, $(i = 1, \ldots, n)$, may be iteratively constructed; where

$$u_i^{(q)} \text{ is } N(0, v_i(\underline{t})) \text{ for } q=1, \ldots, p \text{ and } i=1, \ldots, (n-1)$$

and

$$u_n^{(q)} \text{ is } N(x_0^{(q)}, v_n(\underline{t})) \text{ for } q=1, \ldots, p,$$

$$\left. \right\} \quad (5.5.2)$$

and all variables are independent.

As in Chapter 4 we write

$$\underline{\underline{u}} = \{u_i^{(q)}, q=1, \ldots, p, i=1, \ldots, n\} = \{\underline{u}_i, i=1, \ldots, n\} = \{\underline{u}^{(q)}, q=1, \ldots, p\}.$$
$$(5.5.3)$$

$\underline{u}^{(q)} = \underline{u}(\underline{x}^{(q)}; \underline{t}) = (u_i(\underline{x}^{(q)}; \underline{t}), i = 1, \ldots, n)$ is linear in $\underline{x}^{(q)}$ and the Jacobian of the transformation is 1, for all $\underline{t}$ and every q (Felsenstein (1973); his 't' being a simple transformation of ours). Thus

$$L(\underline{x}_0, \underline{t}, F) = f_{\underline{\underline{X}}}(\underline{\underline{x}} | \underline{x}_0, \underline{t}, F) = f_{\underline{\underline{U}}}(\underline{\underline{u}} | \underline{x}_0, \underline{t}, F),$$

where $f_{\underline{\underline{X}}}$ and $f_{\underline{\underline{U}}}$ denote density functions corresponding to sets of variables $\underline{\underline{x}}$ and $\underline{\underline{u}}$ respectively, and

$$-2S(\underline{x}_0, \underline{t}, F) = p \sum_{i=1}^{n} \log(v_i(\underline{t}))$$
$$+ \sum_{i=1}^{n-1} \{ [\sum_{q=1}^{p} (u_i(\underline{x}^{(q)}; \underline{t}))^2] / v_i(\underline{t}) \}$$
$$+ \sum_{q=1}^{p} (u_n(\underline{x}^{(q)}; \underline{t}) - x_0^{(q)})^2 / v_n(\underline{t}). \quad (5.5.4)$$

Thus

$$\hat{\underline{x}}_0(\underline{\underline{x}}, \underline{t}, F) = (u_n(\underline{x}^{(q)}; t); \quad q = 1, \ldots, p). \quad (5.5.5)$$

Let $(H_i(\underline{\underline{x}}; \underline{t}))^2 = \sum_{q=1}^{p} (u_i(\underline{x}^{(q)}; \underline{t}))^2$, [not to be confused with the function H (or functions D_k) of Chapter 4]; then

118

$$-2S^*(\underline{t}, F) = -2 \log_e L^*(\underline{t}, F)$$

$$= p \sum_{i=1}^{n} \log(v_i(\underline{t})) + \sum_{i=1}^{n-1} ((H_i(\underline{x}; \underline{t}))^2 / v_i(\underline{t})). \quad (5.5.6)$$

Felsenstein (1973) gives a method for the rapid evaluation of H_i^2 and v_i ($i = 1, \ldots, (n-1)$) and v_n, and hence of (5.5.1), for given $(\underline{t}, F)$ and pairwise population distances. Choice of a suitable series of evaluation points $(\underline{t}, F)$ can lead to the determination of a local maximum in the multidimensional space

$$R_+^{(n-1)} \times \{ F_j ; j = 1, \ldots, (n! \, (n-1)! \, / 2^{n-1}) \}.$$

There may be many such local maxima and a search procedure based only on evaluation cannot indicate whether this is the case, or demonstrate the general form of the support surface; the importance of Felsenstein's transformations are that the form (5.5.4) suggests a procedure for the elimination of $\underline{x}_0$, in situations where it may be regarded as a nuisance parameter. For Felsenstein (1973) suggests that $\underline{u}_n$ contains no information about $(\underline{t}, F)$, and that inferences may be based on $\underline{u}_1, \ldots, \underline{u}_{n-1}$ alone. We have then the marginal likelihood (Kalbfleisch and Sprott (1970));

$$L^{**}(\underline{t}, F) = f_{\underline{U}}(\underline{u}_1, \ldots, \underline{u}_{n-1} | \underline{t}, F) \quad (5.5.7)$$

$$-2 \log L^{**}(\underline{t}, F) = p \sum_{i=1}^{n-1} \log v_i(\underline{t}) + \sum_{i=1}^{n-1} (H_i(\underline{x}; \underline{t}))^2 / v_i(\underline{t}) \quad (5.5.8)$$

or

$$-2S^{**}(\underline{t}, F) = -2S^*(\underline{t}, F) - p \log v_n(\underline{t}) \quad (5.5.9)$$

and this may also be rapidly evaluated.

Since $L^{**}(\underline{t}, F)$ is the likelihood used by Felsenstein (1973) we refer to it as the Felsenstein Likelihood (FL) and to the values $(\underline{t}^{**}, F^{**})$ maximising L^{**} as the FL estimates.

The justification offered for the adoption of (5.5.8) has been that it does not have the tendency of the MRL to produce 4-way roots (4.5(ii)). That a model does not produce the required results should be a criticism of the model or of the preconceived results, rather than of the method of inference. The acceptance of (5.5.8) requires further justification, which may be based on the concept of M-ancillarity introduced by Barndorff-Nielsen (1971).

The u_i are independent $(i = 1, \ldots, n)$, and for any given u_{-n} and $(\underline{t}, F)$ there is a value of $\underline{x}_0$ (that is, $\underline{x}_0 = u_{-n}$) such that

$$f_{\underline{U}}(u_{-n}|\underline{x}_0, \underline{t}, F) > f_{\underline{U}}(u^*_{-n}|\underline{x}_0, \underline{t}, F) \text{ for all } u^*_{-n} \neq u_{-n}. \qquad (5.5.10)$$

Further the domains of variation of $\underline{x}_0$ and $(\underline{t}, F)$ are independent. But these are precisely the conditions for M-ancillarity which is based on the concept of universality, as opposed to the more classical concepts of B- and S-ancillarity which are based on a factorisation of the likelihood function.

The statement of M-ancillarity is that: if, whatever happens (u_{-n}) and whatever the values of the parameters of interest $(\underline{t}, F)$, there is a value of the nuisance parameter $(\underline{x}_0)$ that makes what has happened the most probable event, then what has happened is uninformative about the parameters of interest in the absence of further information regarding the nuisance parameter.

There is a further problem in that the u_i are functions of the parameters $\underline{t}$. Following Kalbfleisch and Sprott (1970) we have

$$
\begin{aligned}
f_{\underline{X}}(\underline{x}|\underline{x}_0, \underline{t}, F)d\underline{x} &= f_{\underline{U}}(\underline{u}|\underline{x}_0, \underline{t}, F)d\underline{u} \\
&= (f_{\underline{U}}(u_{-1}, \ldots, u_{-n-1}|\underline{x}_0, \underline{t}, F)du_{-1}\ldots du_{-n-1}) \\
&\quad (f_{\underline{U}}(u_{-n}|u_{-1}, \ldots, u_{-n-1}, \underline{x}_0, \underline{t}, F)du_{-n}) \\
&= (f_{\underline{U}}(u_{-1}, \ldots, u_{-n-1}|\underline{t}, F)du_{-1}\ldots du_{-n-1}) \\
&\quad (f_{\underline{U}}(u_{-n}|\underline{x}_0, \underline{t}, F)du_{-n}).
\end{aligned}
$$

The second term is not independent of $(\underline{t}, F)$, but provided the concept of M-ancillarity is accepted we see from (5.5.10) that it may be deemed to be uninformative about $(\underline{t}, F)$ in the absence of knowledge of $\underline{x}_0$, and inferences regarding $(\underline{t}, F)$ may be based on the first term alone. [To do so must however affect our inferences regarding $(\underline{t}, F)$, since the second term does contain information jointly on $(\underline{x}_0, \underline{t}, F)$.] To further conclude that inferences may be based on the density function (5.5.7) it is not sufficient that the determinant of the Jacobian $(|J|)$ of the transformation, and hence $du_{-1} \ldots du_{-n}$, be independent of t; the subspace volume element $du_{-1} \ldots du_{-n-1}$ must be so also.

From Kalbfleisch and Sprott (1970) we have that

$$\prod_{i=1}^{n-1} du_{-i} = |K'K|^{\frac{1}{2}p} \prod_{i=1}^{n} dx_{-i} / |J|^p,$$

where K is the column vector $(\delta x_i^{(q)} / \delta u_n^{(q)}, i = 1, \ldots, n)$, which is independent of q.

Lemma. $|K'K|$ is independent of t.

Proof. $\underline{u}^{(q)} = J \underline{x}^{(q)} \quad (q = 1, \ldots, p),$ \hfill (5.5.11)

where J is the Jacobian of the linear transformation. Thus $\underline{x}^{(q)} = J^{-1} \underline{u}^{(q)}$ and $|K'K| = \sum_{i=1}^{n} (J^{in})^2 . [J^{in} = (J^{-1})_{in}].$ Taking expectations

$$E(x_i^{(q)}) = \sum_{\ell=1}^{n} J^{i\ell} E(u_\ell^{(q)}),$$ \hfill (5.5.12)

or

$$x_0^{(q)} = J^{in} x_0^{(q)} \text{ for all } (\underline{x}_0, \underline{t}, F),$$

since $E(u_i^{(q)}) = 0$ for $i \neq n$ and $E(u_n^{(q)}) = x_0^{(q)}$, and $E(x_i^{(q)}) = x_0^{(q)}$ for all i.

Thus $J^{in} = 1 \ (i = 1, \ldots, n)$, and $|K'K| = n$ which is independent of $\underline{t}$. //

Thus $\prod_{i=1}^{n-1} du_{-i} = n^{\frac{1}{2}p} \prod_{i=1}^{n} dx_{-i}$ and is independent of $\underline{t}$ $(|J| = 1)$; the Kalbfleisch and Sprott criterion is satisfied, and inferences may be based upon (5.5.8).

We compare now the properties of (5.5.1) and (5.5.8). From (5.5.5) and (4.3.6)

$$\underline{u}_n = \hat{\underline{x}}_0(\underline{x}, \underline{t}, F) = ((\underline{x}^{(q)},T^{-1}\underline{1})/(\underline{1}'T^{-1}\underline{1}), \quad q = 1, \ldots, p)$$

and thus

$$v_n(\underline{t}) = \text{var}(u_n^{(q)}) = (\underline{1}'T^{-1}TT^{-1}\underline{1})/(\underline{1}'T^{-1}\underline{1})^2 = (\underline{1}'T^{-1}\underline{1})^{-1}. \quad (5.5.13)$$

Then from (5. 5. 9)

$$S^{**}(\underline{t}, \ F) = S^*(\underline{t}, \ F) - \tfrac{1}{2}p\log(\underline{1}'T^{-1}\underline{1}). \quad (5.5.14)$$

Using the iterative formula (4. 7. 11) we may show by induction that

$$\frac{\delta}{\delta t_r}(\underline{1}'T^{-1}\underline{1}) < 0 \ \text{ for all } \ \underline{t} > 0, \ r = 1, \ \ldots, \ (n-1).$$

Thus

$$\frac{\delta S^{**}}{\delta t_r} > \frac{\delta S^*}{\delta t_r} \ \text{ for each } \ r \ \text{ and for all } \ \underline{t} > 0. \quad (5.5.15)$$

Also, for each k, S^* and S^{**} are unimodal in t_k, the other t_i remaining fixed, or else decrease monotonically from $t_k = 0$ (4. 5(iii)). Thus, although it is not necessary that each t_i^{**} be individually greater than the corresponding $\hat{t}_i$, (5.5.15) shows that the FL will tend to give larger time interval estimates than the MRL, and hence more bifurcating splits.

However, besides the perhaps desirable tendency to produce fewer multifurcating nodes, the FL has also the property that it usually has internal maxima for several tree forms: simple examples for $n = 3$ and 4 may be readily constructed (5. 6). Computationally the latter property, which apparently does not occur with the MRL, is undesirable: the determination of a local maximum is no longer a sufficient criterion for the acceptance of a tree form. The multifurcating root given by the MRL and the many internal maxima of the FL are both expressions of ignorance as to the true tree form, but the former is necessarily determined while the latter may not be found.

From (5. 5. 9) and (5. 5. 13) it may be shown that

$$L^{**}(\underline{t}, \ F) = \int \ldots \int L(\underline{x}_0, \ \underline{t}, \ F)d\underline{x}_0 = \int \ldots \int f(\underline{x}|\underline{x}_0, \ \underline{t}, \ F)d\underline{x}_0.$$

Thus the FL has the same functional form as a likelihood for $(\underline{t}, \ F)$ when either integration over a Bayesian uniform prior distribution for $\underline{x}_0$, or over the fiducial distribution induced on $\underline{x}_0$ by the data, is acceptable. However the fiducial and Bayesian interpretations of L^{**} may be logically different from its interpretation as a marginal likelihood.

The FL tree is not an unrooted tree. The probability model is for a rooted tree, and, using the FL, we obtain an estimate of the time of this root, even though we cannot estimate its position. Using the FL it is technically possible to relax the requirement that the times of data points be known, in that 'undirected times of divergence' (but not the time of the root) may still be estimated (Felsenstein: personal communication). The 'tree' then estimated would be truly unrooted, but the time intervals could not be interpreted as having evolutionary direction, and the 'tree' could not be translated into an inference of the evolutionary history. It would be a fit of the 'times' of arcs to pairwise population (distances)[2] essentially analogous to the LSA heuristic method (1.4), although more justifiable on the basis of a model of independent increments in that it is the squared distances that are assumed additive. The fact that for data points at variable times the (rooted) evolutionary history is essentially unestimable using the FL corresponds to the singularity of the MRL in the same situation (4.6(ii)).

5.6   FINAL COMPARISON OF SOLUTIONS IN SOME SPECIAL CASES

Chapters 2 to 5 have considered in some detail the estimation of an evolutionary tree from contemporary genetic data. We have considered the process of random genetic drift, the probability model, and the general form of the likelihood. We have seen how ML estimates may be made, and considered further problems arising from this solution. The iterative method of Chapter 4 rapidly estimates a tree, according to a logically justifiable statistical method, on the basis of a probability model which is shown to be a close approximation to a genetic process which is known to be taking place.

For reasons previously discussed we have advocated the use of the MRL, but as shown in the previous section the FL may also have some justification. Besides the ML and FL solutions there are also the heuristic methods of solution, ME and LSA, which are as yet the only methods widely used in practice. In this final section we compare the ML, FL and ME solutions in some special cases. These demonstrate the tendency of FL to have many internal roots, and show that the ME and LSA solutions cannot be considered adequate estimates of an evolutionary tree.

We consider first the case $n = 3$. The ML and ME solutions have been given in 3.3; we retain the notation of that section;

$$
\left.
\begin{aligned}
F_3 &= ((\underline{x}_1, \, \underline{x}_2), \, \underline{x}_3) \\
d^2 &= \| \underline{x}_1 - \underline{x}_2 \|^2 \\
\text{and} \quad \underline{h} &= (\underline{x}_3 - \tfrac{1}{2}(\underline{x}_1 + \underline{x}_2)), \; h^2 = \underline{h} \cdot \underline{h}.
\end{aligned}
\right\}
\tag{5.6.1}
$$

The MRL has an internal root for $F = F_3$ only if $h > 3d/2$ or

$$
\| \underline{x}_3 - \underline{x}_1 \|^2 + \| \underline{x}_3 - \underline{x}_2 \|^2 > 5 \| \underline{x}_1 - \underline{x}_2 \|^2
\tag{5.6.2}
$$

For the FL we have; $u_1^{(q)} = (x_1^{(q)} - x_2^{(q)})$ is $N(0, \, 2t_1)$ and $u_2^{(q)} = h^{(q)}$ is $N(0, \, \tfrac{1}{2}(3t_1 + 4t_2))$, and $\mathrm{cov}(u_1^{(q)}, u_2^{(q)}) = 0$ $q = 1, \ldots, p$. (See Felsenstein (1968) for the method of constructing vectors $\underline{u}_i$.)

Thus $H_1^2 = d^2$ and $H_2^2 = h^2$, $v_1(t) = 2t_1$ and $v_2(t) = \tfrac{1}{2}(3t_1 + 4t_2)$, in the notation of 5.5. Hence (5.5.8) has an internal root for $F = F_3$ provided

$$
4h^2 > 3d^2
$$

or

$$
\| \underline{x}_3 - \underline{x}_2 \|^2 + \| \underline{x}_3 - \underline{x}_1 \|^2 > 2 \| \underline{x}_1 - \underline{x}_2 \|^2,
\tag{5.6.3}
$$

in which case

$$
t_1^{**} = d^2/2p, \quad t_2^{**} = (4h^2 - 3d^2)/8p.
\tag{5.6.4}
$$

Whereas there can never be two tree forms satisfying the condition (5.6.2) (Fig. 3.3(c)), there may often be two satisfying (5.6.3), (Fig. 5.6(a)). The tree form with maximum $L(t^{**}(F), F)$ is the same as that inferred using the MRL.

For the FL solution $\underline{x}_0$ and all internal nodes are not to be considered; acceptance of any $\underline{x}_0$ entails joint estimation and the ML solution. To enable us to specify a tree completely we may however consider

$$
\underline{x}_0^{**} = \hat{\underline{x}}_0(\underline{x}, \, \underline{t}^{**}, \, F^{**}),
\tag{5.6.5}
$$

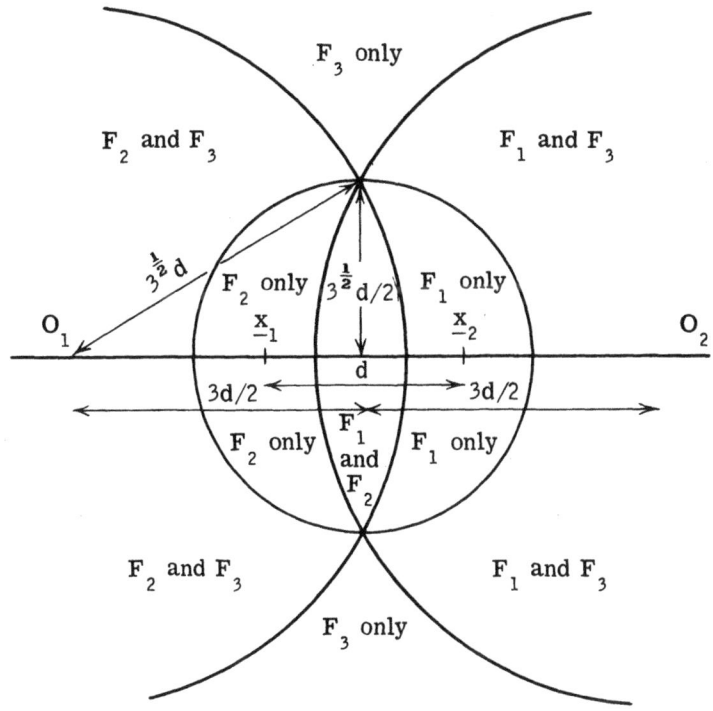

Fig. 5.6(a). The three population case. Diagram showing the tree forms with strictly positive FL estimates of time intervals as $\underline{x}_3$ varies over the plane with $\underline{x}_1$ and $\underline{x}_2$ remaining fixed. Compare this with Figs. 3.3(c) and 3.3(d).

and mean internal node positions given $(\underline{x}_0^{**}, \underline{t}^{**}, F^{**})$ and $\underline{x}$. The results in this case are of similar form to those for the MRL given in 3.3(ii).

For $n = 4$ we consider

$$\underline{x}_1 = (0, 0, 0, \ldots, 0), \quad \underline{x}_2 = (0, y, 0, \ldots, 0),$$
$$\underline{x}_3 = (x, 0, \ldots, 0) \text{ and } \underline{x}_4 = (x, y, 0, \ldots, 0),$$

with all other notation as given by Fig. 5.6(b).

125

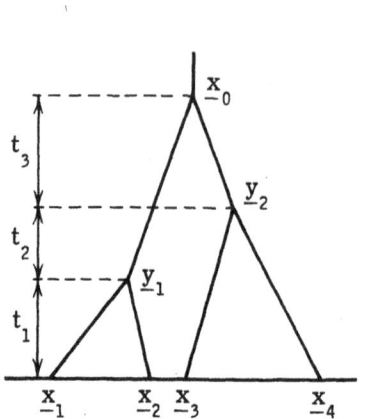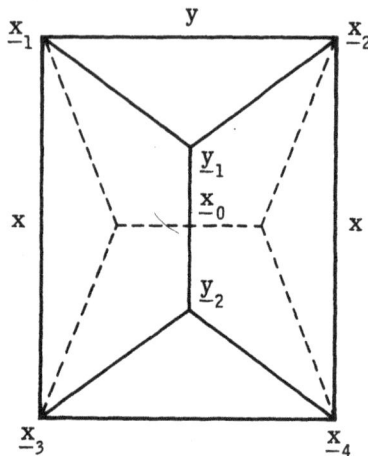

Fig. 5.6(b).   The case of four populations situated in a rectangle.
The case $F = F_1$ is shown, and the broken line
indicates how a tree with internal maximum could
also be inferred for $F = F_2$.

Let $F_1 = ((\underline{x}_1, \underline{x}_2), (\underline{x}_3, \underline{x}_4))$ and $F_2 = ((\underline{x}_1, \underline{x}_3), (\underline{x}_2, \underline{x}_4))$.
For the MRL with $F = F_1$ we have, if $x > y$,

$$
\left.
\begin{aligned}
&\hat{\underline{x}}_0 = (\tfrac{1}{2}x, \tfrac{1}{2}y, 0, \ldots, 0) \\
&\hat{t}_1 = y^2/2p, \ \hat{t}_2 = 0, \ \hat{t}_3 = (x^2 - y^2)/4p,
\end{aligned}
\right\}
\qquad (5.6.6)
$$

then

$$
E(\underline{y}_1 \,|\, \underline{\underline{x}}, \underline{x}_0, \underline{t}, F_1) = (y^2/2x, \tfrac{1}{2}y, 0, \ldots, 0) \quad \text{and}
$$
$$
E(\underline{y}_1 + \underline{y}_2 \,|\, \underline{\underline{x}}, \underline{x}_0, \underline{t}, F_1) = (x, y, 0, \ldots, 0). \qquad (5.6.7)
$$

There is an internal stationary point for $F = F_1$ only if $x > y$;
the point is then unique and gives a maximum of the MRL. There can thus
never be internal stationary points for both F_1 and F_2. [We refer to
maxima with $t_1 > 0$, $t_3 > 0$ as 'internal' in this case, although $\hat{t}_2 = 0$
by symmetry.] $(5.6.8)$
    Neither does the alternative topology give trees with internal
maxima.

126

For the FL it may be shown that $t_2 = 0$ at any stationary point. Then for $F = F_1$;

$$u_1^{(q)} = (x_1^{(q)} - x_2^{(q)}) \qquad\qquad \text{is } N(0, 2t_1)$$

$$u_2^{(q)} = (x_3^{(q)} - x_4^{(q)}) \qquad\qquad \text{is } N(0, 2t_1)$$

and $\quad u_3^{(q)} = \tfrac{1}{2}(x_3^{(q)} + x_4^{(q)}) - \tfrac{1}{2}(x_1^{(q)} + x_2^{(q)}) \text{ is } N(0, (t_1 + 2t_3))$,

all three vectors being jointly independent. [$\underline{u}$ remains independent of $\underline{t}$ for this topology.] Then $H_1^2 = H_2^2 = y^2$ and $H_3^2 = x^2$, and $v_1(\underline{t}) = v_2(\underline{t}) = 2t_1$ and $v_3(\underline{t}) = t_1 + 2t_3$. Thus if $2x^2 > y^2$, (5.5.8) has a unique internal stationary point for $F = F_1$, this being a maximum of S^{**}, and

$$t_1^{**} = (y^2/2p), \quad t_2^{**} = 0 \quad \text{and} \quad t_3^{**} = (2x^2 - y^2)/4p. \qquad (5.6.9)$$

$$\left.\begin{array}{l}
\underline{x}_0^{**} = (\tfrac{1}{2}x, \tfrac{1}{2}y, 0, \ldots, 0) \\[4pt]
E(\underline{y}_1 | \underline{x}, \underline{x}_0^{**}, \underline{t}^{**}, F_1) = (y^2/4x, \tfrac{1}{2}y, 0, \ldots, 0) \\[4pt]
E(\underline{y}_1 + \underline{y}_2 | \underline{x}, \underline{x}_0^{**}, \underline{t}^{**}, F_1) = (x, y, 0, \ldots, 0)
\end{array}\right\} \qquad (5.6.10)$$

and

Note that:

(i) $\quad t_1^{**} = \hat{t}_1, \; t_3^{**} > \hat{t}_3$; FL estimates are larger than ML.

(ii) If $y^2 < 2x^2 < 4y^2$ both F_1 and F_2 give $t_3^{**} > 0$. (5.6.11) This cannot occur with the MRL.

(iii) $\quad L^{**}(\underline{t}^{**}(F_1), F_1) > L^{**}(F_2), F_2)$ if and only if $x > y$.

(iv) This case is not 'pathological': distortions of the symmetrical case give situations in which

$(t_1^{**}, t_2^{**}, t_3^{**})$ is strictly positive for $F = F_1$ and $F = F_2$.

ME gives a Steiner tree for the above data and $F = F_1$ if

$$x > y/3^{\frac{1}{2}}.$$

F_1 and F_2 are both Steiner if $y^2 < 3x^2 < 9y^2$. $\qquad\qquad (5.6.12)$

If $x > y$ F_1 has the shorter total length.

The internal nodes for F_1 are given by

$$\underline{y}_1 = (\tfrac{1}{2}y/3^{\frac{1}{2}}, \ \tfrac{1}{2}y, \ 0, \ \ldots, \ 0) \quad \text{and} \quad (\underline{y}_1 + \underline{y}_2) = (x, \ y, \ 0, \ \ldots, \ 0).$$

$$(5.\,6.\,13)$$

Thus although if $x > y$ F_1 is in all cases preferred to F_2 comparison of (5. 6. 8), (5. 6. 11) and (5. 6. 12) and of (5. 6. 7), (5. 6. 10) and (5. 6. 13) show that the details are very different.

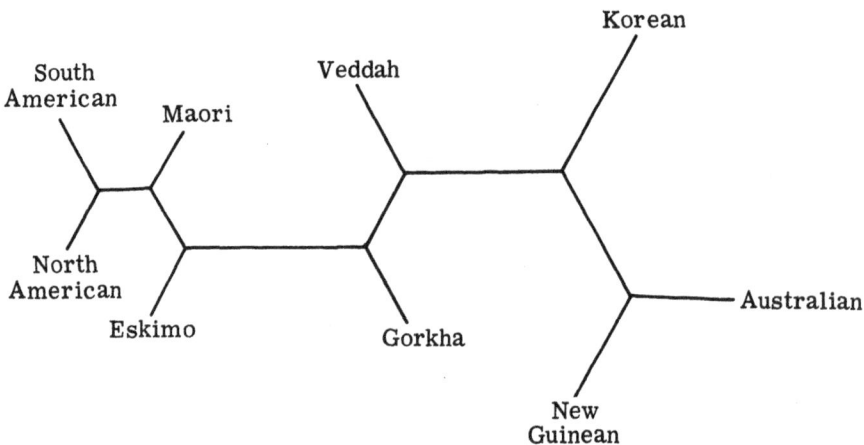

Fig. 5. 6(c).   The form of the ME tree inferred from the A and A data of section 5. 1.

We compare finally the ME solutions for the two data sets of 5. 1 with the ML solutions given there.   The ME algorithms have been tested extensively on the A and A data (Thompson (1973a)), and many Steiner trees have been found.   The shortest is that produced by the Prim method (Fig. 5. 6(c)); this agrees with the ML solution in distinguishing the two major ethnic groups, but differs significantly in detail.   The ML tree with a 4-way root corresponds to any of three unrooted trees; one of these is Steiner but is longer than that of Fig. 5. 6(c), which conversely has low likelihood.

For the N. W. Eu. data the unrooted tree corresponding to the ML form is Steiner and has length 0. 413.   The form with Norway and Sweden reversed is also Steiner and has length 0. 411 being the shortest found. However this form has very low support.   Thus while ML trees are frequently good ME trees, the best ME trees are often of very low likelihood. Fig. 5.6(d) shows why this is so, with reference to the above case of the N.W. Eu. data. Although the total length of the arcs shown are virtually equal in the two cases, the length of the terminal arcs of the second tree are compatible with contemporary populations; those of the first are not. A likelihood criterion, which includes contemporary data as part of the probability model, distinguishes these trees, whereas ME cannot, since it has no time structure.

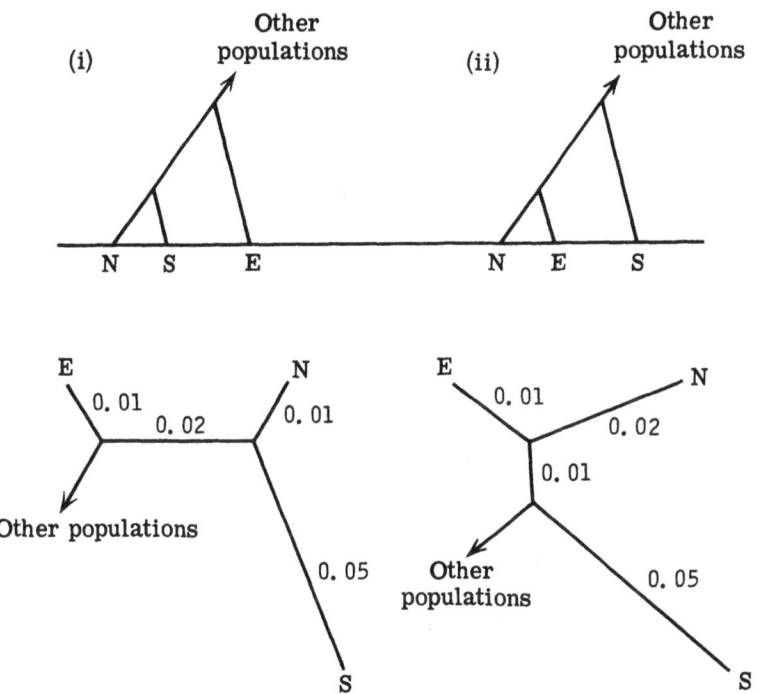

Fig. 5.6(d).   The two alternative forms for the subtree of Norway (N), Sweden (S) and England (E) inferred from the N.W. Eu. data of section 5.1. ME criteria are unable to distinguish the two forms, but, whereas tree (ii) is the ML tree, tree (i) has very low support.

Thus ME, and equally LSA, do not provide good estimates of the evolutionary tree, not only because they produce unrooted trees and because of qualitative differences in the positions of internal nodes, but more seriously because having no time structure they have no criterion of time compatibility of arcs. It is thus surprising that ME, which is a non-hierarchic cluster method, usually produces adequate representations of the projection into the present of the hierarchic ML evolutionary tree. This can only be because, in the particular cases in which these methods have been successful, the phylogenetic relationships are sufficiently well defined to be determined by any objective criterion.

6·The Icelandic admixture problem

6.1 INTRODUCTION

In this final chapter we consider a specific admixture problem to which the general model of independently evolving populations, subject to forces of random genetic drift, may be applied. This is the problem of estimating the proportions in the Norse-Celtic mixture which formed the colonising population of Iceland.

Iceland was colonised by Norsemen between A. D. 874 and A. D. 930. By 950 the population was 50,000 and remained between 50 and 70 thousand until 1900. This constancy of population size, and the accuracy with which it is known through the very long tradition of national censuses, makes the Icelandic population particularly convenient for study. It has been estimated (J. H. Edwards; personal communication) that the harmonic mean of the population size since A. D. 950 has been 60,000, and thus the effective population size has been around 30,000 (2.1).

Many of the Norse colonists had spent some time in Ireland or Western Scotland before colonising Iceland, and many may have had Irish slaves. Thus the Icelandic population of A. D. 950 was a Celtic-Norse mixture, in the proportions $(1 - r):r$ say. The aim is to estimate r $(0 \leq r \leq 1)$ from present-day gene frequencies. It used to be thought that the population was predominantly Norse. This is what the sagas claim but these were written long after colonisation and there is virtually no contemporary evidence. The language and culture are Norse and all later political links were with Scandinavia. Although this indicates that the ruling classes, at least, were Norse, studies of blood group gene frequencies indicate a considerable Celtic component in the population. This was first noted by Donegani et al. (1950) with some small samples of ABO, Rhesus and MN data. More recent studies (Constandse-Westermann (1972), Bjarnason et al. (1973)) have amply confirmed their suspicions. We have

131

already seen (5.1) that an evolutionary tree model gives a Celtic origin for the Icelanders.

We now investigate the problem via a new model of a Celtic-Norse mixture in A.D. 950 followed by 1000 years (40 generations) of random genetic drift modifying the gene frequencies of the Celts, becoming Irish, Norse, becoming Norwegians, and Icelanders (Fig. 6.2(a)).

The major assumptions of the model are that it has been random genetic drift that has influenced the population gene frequencies of Icelanders, Irish and Norwegians over the last 1000 years, and that the individuals sampled now are representative descendants of the relevant Celtic, Norse and Icelandic populations. The main factors that could invalidate this are, firstly, differential selection, secondly, migration to any of the three countries subsequent to the original mixture, and thirdly non-representative sampling. This last could take the form either of the original Norsemen having originated from an atypical region of Norway, or of present Icelandic samples coming from regions with a particularly large Irish component. Fuller discussion of these factors is given by Thompson (1973b), but the conclusion is that in the case of the Icelanders a model of independently evolving populations and random genetic drift is appropriate. However any, or all, of the above factors may make the model invalid for other admixture problems.

6.2 THE MODEL

As in the previous chapters we transform the observed gene frequencies to give vectors in a p-dimensional Euclidean space, in which the process of random genetic drift becomes one of approximate Brownian motion. Suppose that the present observed sample frequencies of Norwegians, Irish and Icelanders give vectors

$$\underline{x}_n, \ \underline{x}_c \ \text{and} \ \underline{x}_i,$$

where $\underline{x}_n = (x_n^{(q)}, q = 1, \ldots, p)$, while the present true population frequencies give

$$\underline{x}_n^*, \ \underline{x}_c^* \ \text{and} \ \underline{x}_i^* .$$

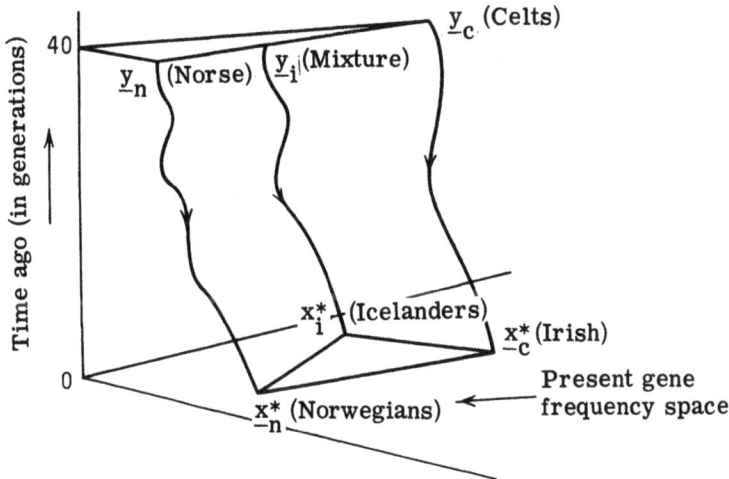

Fig. 6.2(a).   Diagrammatic representation of the mathematical model. Norse, Celts and a direct mixture thereof, 40 generations ago, have gene frequencies which change under the process of random genetic drift to those of present-day Norwegians, Irish and Icelanders. $\underline{y}_i = r\underline{y}_n + (1-r)\underline{y}_c$ for some r, $0 \le r \le 1$. [Diagram from Thompson (1973b).]

Suppose further than the unknown initial frequencies t generations ago (t is approximately 40) of Norsemen, Celts and the colonising mixture give vectors $\underline{y}_n$, $\underline{y}_c$ and $\underline{y}_i$ (see Fig. 6.2(a)).

Now it is assumed that the original Icelanders were a simple mixture of Celts and Norsemen. Thus if the gene frequencies at some k-allele locus were $p_n^{(j)}$, $p_c^{(j)}$ and $p_i^{(j)}$ for $j = 1, \ldots, k$, then

$$p_i^{(j)} = (1-r)p_c^{(j)} + rp_n^{(j)} = p_c^{(j)} + r(p_n^{(j)} - p_c^{(j)}), \quad j = 1, \ldots, k,$$

the sampling involved in forming the mixture being equivalent to one generation of drift. It may be simply shown that these equations transform to

$$y_i^{(q)} = y_c^{(q)} + r(y_n^{(q)} - y_c^{(q)}) + 0((r\|\underline{y}_n - \underline{y}_c\|)^2), \quad q = 1, \ldots, p,$$

133

and hence, over the small region of the projected space concerned,

$$\underline{y}_i = r\underline{y}_n + (1-r)\underline{y}_c \text{ to an accurate approximation.}$$

Finally suppose that the effective population sizes over the period since A. D. 950 are N_n, N_c and N_i and the sample sizes, assumed equal for all loci, are m_n, m_c and m_i.

Then we have the Normal approximations: as in section 3.1

$$x_n^{(q)}* \text{ is } N(y_n^{(q)}, t/8N_n) \text{ for } q = 1, \ldots, p$$

and as in 5.3

$$x_n^{(q)} \text{ is } N(x_n^{(q)}*, 1/8m_n) \text{ for } q = 1, \ldots, p,$$

this latter equation holding strictly only if all genotypes are identifiable, but as a reasonable approximation provided the number of phenotypes exceeds the number of alleles. Hence

$$x_n^{(q)} \text{ is } N(y_n^{(q)}, (t/8N_n)+(1/8m_n))=N(y_n^{(q)}, \sigma_n^2) \text{ } q=1, \ldots, p.$$

Similarly

$$x_c^{(q)} \text{ is } N(y_c^{(q)}, (t/8N_c)+(1/m_c))=N(y_c^{(q)}, \sigma_c^2) \text{ } q=1, \ldots, p,$$

and

$$x_i^{(q)} \text{ is } N(y_i^{(q)}, (t/8N_i)+(1/8m_i))=N(y_i^{(q)}, \sigma_i^2) q=1, \ldots, p,$$

(6.2.1)

all components being independent and

$$y_i^{(q)} = r y_n^{(q)} + (1 - r) y_c^{(q)}, \quad q = 1, \ldots, p. \tag{6.2.2}$$

If sample sizes vary between loci not all the components of each population vector have the same variance. This complicates the analysis but does not essentially alter the situation. We note that a negative estimate of r could be interpreted as an inference that the Irish are a Norwegian-Icelandic mixture, and $r > 1$ as the Norwegians being an Icelandic-Irish mixture, but unless these are hypotheses which we are a priori prepared to consider we may restrict attention to the support function within the range $0 \le r \le 1$.

The problem of estimating the mixture proportions in a hybrid population was considered by Glass and Li (1953), and their solution has been elaborated by Krieger et al. (1965) and Elston (1971). The two basic criteria of estimation that have been used are maximum likelihood and least-squares, but although the estimation methods are sophisticated and can take into account dominance at some gene loci and mixtures of several populations, the models have all contained the same basic assumptions. These are that the population frequencies in the unmixed populations are known and are the same as in the original ancestral populations, and that the only reason for the observed frequencies not being a simple mixture at all loci is sampling in the hybrid population $(\underline{x}_n = \underline{x}_n^* = \underline{y}_n,$ $\underline{x}_c = \underline{x}_c^* = \underline{y}_c, \underline{x}_i^* = \underline{y}_i)$. Although this may be appropriate for very recent mixtures where the parent populations have been extensively surveyed, it will not be so in a situation where the hybrid population has been separated from any of its contributors for any length of time. A simple sampling model is not justifiable in many of the situations to which it has been applied, and although the present model involves several approximations it is an approximation based on the true major causes of observed gene frequency differentiation.

The simple sampling model may be considered as a special limiting case of (6.2.1) in which N_n, N_c and $N_i \to \infty$ (or $t \to 0$) and m_n and $m_c \to \infty$; σ_n^2 and $\sigma_c^2 \to 0$ and $\sigma_i^2 \to 1/8m_i$. Then

$$x_i^{(q)} \text{ is } N(rx_n^{(q)} + (1-r)x_c^{(q)}, 1/8m_i) \text{ for } q=1, \ldots, p. \qquad (6.2.3)$$

To obtain the ML (or indeed least-squares) estimate, $\hat{r}$, of r we would minimise

$$\sum_{q=1}^{p} (x_i^{(q)} - rx_n^{(q)} - (1-r)x_c^{(q)})^2,$$

obtaining

$$(\underline{x}_i - \hat{r}\underline{x}_n - (1 - \hat{r})\underline{x}_c) \cdot (\underline{x}_n - \underline{x}_c) = 0. \qquad (6.2.4)$$

The sampling model solution is then that shown in Fig. 6.2(b), the distance $\hat{\Pi}$ being a sampling distance.

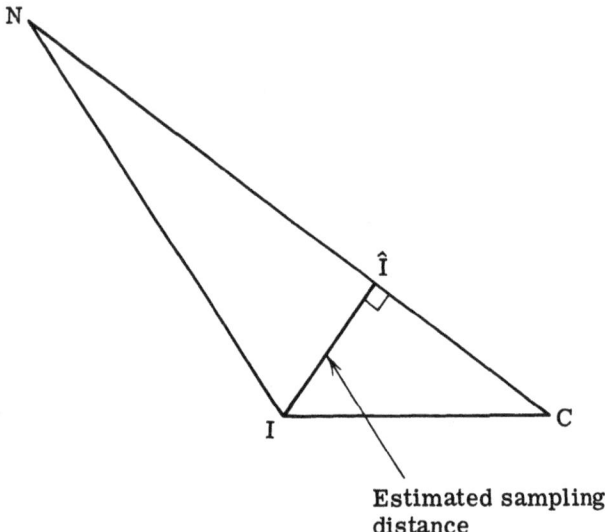

Estimated sampling
distance

Fig. 6.2(b). The sampling model solution. The position vectors $\underline{x}_n$, $\underline{x}_c$, $\underline{x}_i$ and $\hat{\underline{x}}_i^*$ give coplanar points N, C, I and $\hat{I}$ respectively, where $\hat{\underline{x}}_i^* = \hat{r}\underline{x}_n + (1-\hat{r})\underline{x}_c$ and $\hat{r}$ is given by $(\underline{x}_i - \hat{\underline{x}}_i^*).(\underline{x}_n - \underline{x}_c) = 0$. Thus $N\hat{I}$ and $\hat{I}C$ are in the ratio $(1-\hat{r})$ to $\hat{r}$ and $\hat{I}I$ is orthogonal to NC. [Diagram from Thompson (1973(b).]

6.3 THE LIKELIHOOD SOLUTION

Returning to the general drift and sampling model (6.2.1) we now derive a likelihood solution. We have

$$L(r, \underline{y}_n, \underline{y}_c, \sigma_n^2, \sigma_c^2, \sigma_i^2) = f(\underline{x}_n, \underline{x}_c, \underline{x}_i | r, \underline{y}_n, \underline{y}_c, \sigma_n^2, \sigma_c^2, \sigma_i^2)$$

$$\propto (\sigma_n^2 \sigma_c^2 \sigma_i^2)^{-\frac{1}{2}p} \exp[-\tfrac{1}{2}(\sigma_i^{-2} \sum_{q=1}^{p} (x_i^{(q)} - ry_n^{(q)} - (1-r)y_c^{(q)})^2$$

$$+ \sigma_n^{-2} \sum_{q=1}^{p} (x_n^{(q)} - y_n^{(q)})^2 + \sigma_c^{-1} \sum_{q=1}^{p} (x_c^{(q)} - y_c^{(q)})^2)].$$

Hence

$$-2S(r, \underline{y}_n, \underline{y}_c, \sigma_n^2, \sigma_c^2, \sigma_i^2)$$

$$= p \log(\sigma_n^2 \sigma_c^2 \sigma_i^2) + \sigma_i^{-2}(\underline{x}_i - r\underline{y}_n - (1-r)\underline{y}_c).(\underline{x}_i - r\underline{y}_n - (1-r)\underline{y}_c)$$

$$+ \sigma_n^{-2}(\underline{x}_n - \underline{y}_n).(\underline{x}_n - \underline{y}_n) + \sigma_c^{-2}(\underline{x}_c - \underline{y}_c).(\underline{x}_c - \underline{y}_c), \qquad (6.3.1)$$

where $\sigma_i^2 = ((1/8m_i) + (t/8N_i))$ is the total drift and sampling variance of each $x_i^{(q)}$ $(q = 1, \ldots, p)$, etc.

Now although, with some provisos (6.4), it is possible to estimate both the variance $(\sigma_n^2, \sigma_c^2, \sigma_i^2)$ and the parameters $(r, \underline{y}_n, \underline{y}_c)$ from the data $(\underline{x}_n, \underline{x}_c, \underline{x}_i)$, in the present problem it may be assumed that the three variances may be estimated from historical data independently of the genetic information, and hence may be treated here as known constants.

Now let $h(r) = \underline{x}_i - r\underline{x}_n - (1-r)\underline{x}_c$ and $f(r) = r^2\sigma_n^2 + (1-r)^2\sigma_c^2 + \sigma_i^2$.
Note that

$$E(h^{(q)}(r)) = 0 \text{ and } var(h^{(q)}(r)) = f(r) \text{ for each } q, \ 1 \le q \le p. \quad (6.3.2)$$

Maximising (6.3.1) w.r.t. $\underline{y}_n$ and $\underline{y}_c$ we obtain, after some rearrangement,

$$\left.\begin{array}{l} (\hat{\underline{y}}_n - \underline{x}_n)f(\hat{r}) = \hat{r}\sigma_n^2 h(\hat{r}) \\[2mm] (\hat{\underline{y}}_c - \underline{x}_c)f(\hat{r}) = (1 - \hat{r})\sigma_c^2 h(\hat{r}), \end{array}\right\} \qquad (6.3.3)$$

and

where $(\hat{r}, \hat{\underline{y}}_n, \hat{\underline{y}}_c)$ is the joint ML estimate of the parameters given independently estimated variances.

From (6.3.3) we have also

$$(\underline{x}_i - \hat{\underline{y}}_i)f(\hat{r}) = \sigma_i^2 h(\hat{r}), \qquad (6.3.4)$$

where $\hat{\underline{y}}_i = \hat{r}\hat{\underline{y}}_n + (1 - \hat{r})\hat{\underline{y}}_c$.

Maximising (6.3.1) w.r.t. r we further have

$$(\underline{x}_i - \hat{\underline{y}}_i).(\hat{\underline{y}}_n - \hat{\underline{y}}_c) = 0. \qquad (6.3.5)$$

Equations (6.3.3)-(6.3.5) show how the solution may be represented diagrammatically (Fig. 6.3(a)). This diagram may be compared with that giving the solution for the sampling model $(\sigma_n^2, \sigma_c^2 \to 0;$ Fig. 6.2(b)).

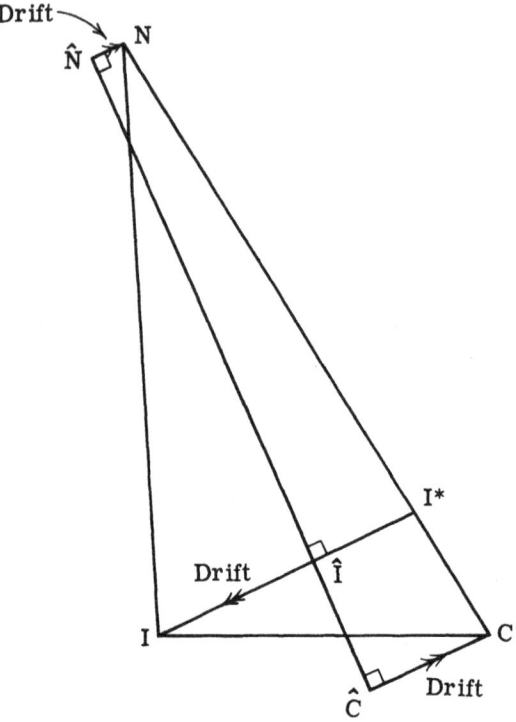

Fig. 6.3(a). The general drift and sampling model solution. Position vectors $\underline{x}_n$, $\underline{x}_c$, $\underline{x}_i$, $\hat{\underline{y}}_n$, $\hat{\underline{y}}_c$, $\hat{\underline{y}}_i$ and $[\hat{r}\underline{x}_n + (1-\hat{r})\underline{x}_c]$ give coplanar points N, C, I, $\hat{N}$, $\hat{C}$, $\hat{I}$ and I* respectively, where $\hat{\underline{y}}_i = \hat{r}\hat{\underline{y}}_n + (1-\hat{r})\hat{\underline{y}}_c$. The three estimated drift vectors $(\hat{\underline{y}}_n - \underline{x}_n)$, $(\hat{\underline{y}}_c - \underline{x}_c)$ and $(\hat{\underline{y}}_i - \underline{x}_i)$ are all orthogonal to the line joining the colinear points $\hat{N}$, $\hat{C}$ and $\hat{I}$. Thus lengths $\hat{N}\hat{I}$ and $\hat{I}\hat{C}$ are in the ratio $(1-\hat{r})$ to $\hat{r}$ as also are NI* and (for equal variances) I*C, and $\hat{C}C$ and $\hat{N}N$, while $\hat{C}C + \hat{N}N + \hat{I}I$. The line II* has the magnitude and direction of the vector $\underline{h}(\hat{r})$.

To solve explicitly for the parameter of interest, r, let d_{in}, d_{nc} and d_{ic} be the pairwise genetic distances (distances in the p-dimensional Euclidean space) between the present observed population positions. Then we have the cosine formulae,

138

$$2(\underline{x}_n - \underline{x}_i) \cdot (\underline{x}_c - \underline{x}_i) = d^2_{ic} + d^2_{in} - d^2_{nc} \quad \text{etc.} \tag{6.3.6}$$

Substituting for $\underline{\hat{y}}_n$ and $\underline{\hat{y}}_c$ from (6.3.3) into (6.3.5) and using (6.3.6) we obtain the quadratic equation,

$$\hat{r}^2[\sigma^2_n(d^2_{in} - d^2_{ic} - d^2_{nc}) + \sigma^2_c(d^2_{in} - d^2_{ic} + d^2_{nc})]$$

$$+ 2\hat{r}[\sigma^2_c(d^2_{ic} - d^2_{nc}) + \sigma^2_n d^2_{ic} - \sigma^2_i d^2_{nc}]$$

$$+ [\sigma^2_i(d^2_{ic} - d^2_{in} + d^2_{nc}) + \sigma^2_c(d^2_{nc} - d^2_{ic} - d^2_{in})] = 0. \tag{6.3.7}$$

This equation always has two distinct real roots except when the co-efficient of $\hat{r}^2$ is zero.

For simplicity we now consider only the case in which all three variances are equal, say to σ^2, although the general case is qualitatively similar. In this case $f(r) = \sigma^2(1 + r^2 + (1-r)^2) = \sigma^2 f^*(r)$ say, and equations (6.3.3)-(6.3.7) reduce to

$$\left. \begin{array}{l} (\underline{\hat{y}}_n - \underline{x}_n) = \hat{r}\underline{h}(\hat{r})/f^*(\hat{r}) \\[2mm] (\underline{\hat{y}}_c - \underline{x}_c) = (1-\hat{r})\underline{h}(\hat{r})/f^*(\hat{r}) \\[2mm] \text{and} \quad (\underline{x}_i - \underline{\hat{y}}_i) = \underline{h}(\hat{r})/f^*(\hat{r}), \end{array} \right\} \tag{6.3.8}$$

and $g(\hat{r}) = 0$, where

$$g(r) = r^2(d^2_{in} - d^2_{ic}) + 2r(d^2_{ic} - d^2_{nc}) + (d^2_{nc} - d^2_{in}), \tag{6.3.9}$$

and $g(r) = 0$ has two distinct real roots unless $d^2_{in} = d^2_{ic}$. Thus the equations for a stationary value of S may be solved, but there are two roots to these equations. It is therefore necessary to reconsider the form of the support function.

Let

$$S^*(r, \sigma^2) = S(r, \underline{\hat{y}}_n(r), \underline{\hat{y}}_c(r), \sigma^2, \sigma^2, \sigma^2)$$

be the maximum relative support function (MRS) for r.

Then from (6.3.1) and (6.3.8)

$$-2S^*(r, \sigma^2) = 3p \log \sigma^2 + h^2(r)/\sigma^2 f^*(r), \qquad (6.3.10)$$

where $h^2(r) = \|\underline{h}(r)\|^2 = \sum_{q=1}^{p} (h^{(q)}(r))^2$.

Using (6.3.6), and if σ^2 is a known constant, we have

$$S^*(r) = -(r^2 d_{nc}^2 - r(d_{ic}^2 + d_{nc}^2 - d_{in}^2) + d_{ic}^2)/[2\sigma^2(1 + r^2 + (1-r)^2)] \quad (6.3.11)$$

and $\dfrac{\delta S^*}{\delta r} = g(r)/[\sigma^2(f^*(r))^2]$.

Thus $\dfrac{\delta S^*}{\delta r}$ has the same sign as $g(r)$ and one of the two roots of $g(r) = 0$ gives a maximum of S^* and the other a minimum. [If $d_{in}^2 = d_{ic}^2 = d^2$ the unique root $r = \frac{1}{2}$ is a maximum/minimum as $d_{nc} \gtrless d$.]

The region of interest is $0 \le r \le 1$,

$$S^*(0) = -d_{ic}^2/4\sigma^2, \quad S^*(1) = -d_{in}^2/4\sigma^2$$

and as $r \to \pm\infty$, $S^*(r) \to -d_{nc}^2/4\sigma^2$.

Thus there are two possible forms for S^* (Fig. 6.3(b)):

(i) $d_{in} > d_{ic}$; $S^*(0) > S^*(1)$, and the coefficient of r^2 in $g(r)$ is positive. The smaller root of $g(r) = 0$ gives the ML estimate of r.

(ii) $d_{ic} > d_{in}$; $S^*(1) > S^*(0)$, and the coefficient of r^2 in $g(r)$ is negative. The larger root of $g(r) = 0$ gives the ML estimate of r.

Since S^*, and indeed the whole problem, is symmetric with respect to interchange of Norse and Celt, n and c, and r and $(1-r)$, it is only necessary to consider the case $d_{in} > d_{ic}$, $(S^*(0) > S^*(1))$. Then

if $d_{in} > d_{nc} > d_{ic}$ $g(0) < 0$, $g(1) < 0$,
         maximum of S^* is in $r < 0$ (and the minimum in $r > 1$),
         ML estimate in $0 \le r \le 1$ is $\hat{r} = 0$.

if $d_{nc} > d_{in} > d_{ic}$ $g(0) > 0$, $g(1) < 0$,
         ML estimate $\hat{r}$ is such that $0 < \hat{r} < 1$.

if $d_{in} > d_{ic} > d_{nc}$ $g(0) < 0$, $g(1) > 0$,
         maximum of S^* is in $r < 0$, minimum in $0 \le r \le 1$. But $S^*(0) > S^*(1)$, thus ML estimate in $0 \le r \le 1$ is $\hat{r} = 0$.

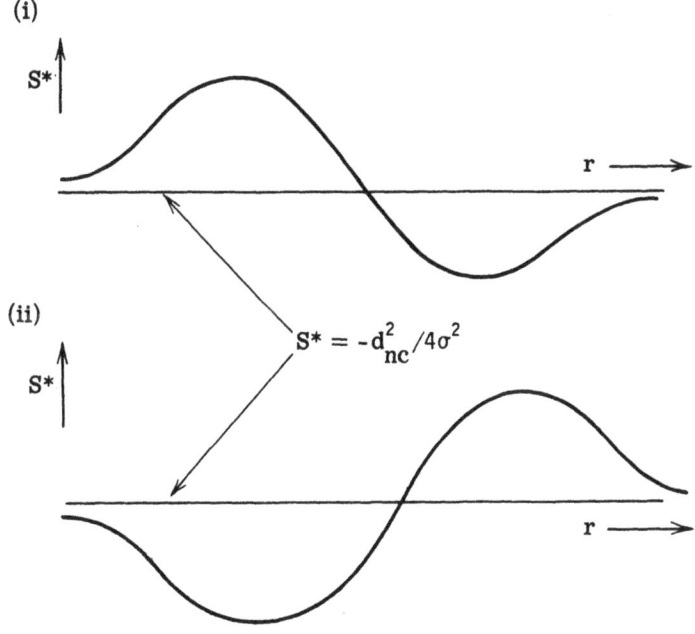

(i)

S*

r

(ii)

S*

$S* = -d_{nc}^2/4\sigma^2$

r

Fig. 6.3(b).   The two possible forms of the function S*.

(i) $d_{in} > d_{ic}$; $S*(0) > S*(1)$. The smaller root of
$g(r) = 0$ gives the ML estimate of r.

(ii) $d_{ic} > d_{in}$; $S*(1) > S*(0)$. The larger root of
$g(r) = 0$ gives the ML estimate of r.

Thus we find that joint maximum likelihood estimation provides
an estimate $\hat{r}$ in the range $0 < \hat{r} < 1$ only if d_{nc} is the largest of the
three pairwise distances. Although this might be expected to be the case
for data arising in admixture situations this restriction will not always
be satisfied. The requirement is far stricter than that for a root in
$0 < r < 1$ under the simple sampling model. Support regions for the
complete set of parameters, $(r, \underline{y}_n, \underline{y}_c)$, may be computed but are not
very meaningful. It is simpler to consider only the MRS, S*, which may
be easily plotted, for given σ^2; two-unit support limits for r may be
found directly from (6.3.11).

We note finally that the sampling model (6.2.3) gives a MRS

$$S^*(r) = -h^2(r)/2(1/8m_i) = -4m_i h^2(r), \qquad (6.3.12)$$

the sample size m_i being simply a scale factor in this case, as is $1/\sigma^2$ in (6.3.11).

6.4 THE DATA AND SOME FURTHER ASPECTS

The relevant present-day descendants of the Norse and Celtic populations which contributed to the Icelandic colonising population are the Norwegians and North-Western Irish. Data on five blood group loci, together giving $p = 6$, give

$$d^2_{ic} = 0.0048, \quad d^2_{in} = 0.0324 \text{ and } d^2_{nc} = 0.0318.$$

[A table of the detailed gene frequency data, provided by Professor J. H. Edwards, is given by Thompson (1973b).]

Under the model of equal variances the roots of $g(r) = 0$ are $r = -0.0053$ (maximum of S^*) and $r = 1.966$ (minimum).

$S^*(0) = -6$ and $S^*(1) = -40.5$, and $d_{in} > d_{ic}$ so that the ML estimate of r (within $0 \leq r \leq 1$) is $\hat{r} = 0$. We therefore estimate that the Icelanders are of wholly Celtic origin and the hypothesis $r = 0$ (wholly Celtic) is $e^{34.5}$ (or 10^{15}) times as likely as the hypothesis $r = 1$ (wholly Norse).

For the Icelanders $N_i \approx 30,000$, $t \approx 40$ and $m_i \approx 3,000$.

Hence σ_i^2 $[= (1/8m_i) + (t/8N_i)]$ is approximately 2×10^{-4}.

Taking this as the population variance for all three populations the support function (6.3.11) may be plotted (Fig. 6.4). The two-unit support limit, given by the solution of $S^*(\hat{r}) - S^*(r) \leq 2$, is $r \leq 0.19$. Thus any value of r up to about 20% falls within the two unit support limit, indicating that any estimate within this range cannot be rejected on these genetic data.

Estimates of the variances from population and sample sizes indicate that, while $\sigma_i^2 = 2 \times 10^{-4}$ is reasonable, a better estimate of σ_n^2 and σ_c^2 is 1×10^{-4} contributed almost entirely by sampling. If we solve the more general equation (6.3.7) with $\sigma_n^2 = \sigma_c^2 = \frac{1}{2}\sigma_i^2$ we obtain $\hat{r} = 0.017$,

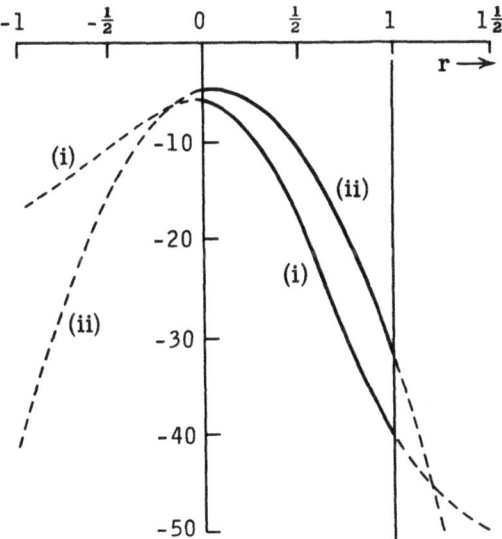

Fig. 6.4. The support function for the admixture fraction r
(0 ≤ r ≤ 1). (i) The support for r in the case of the
general model for the data of Thompson (1973b) and
$\sigma^2 = 2 \times 10^{-4}$, σ^2 being estimated from data on population and sample sizes, independently of the genetic
data, and assumed equal in the three populations.
(ii) The support function for the same maximum likelihood estimates of gene frequency with a classical
sampling model and hypothetical Icelandic sample size
of 250, assumed equal for all loci. [Diagram from
Thompson (1973b).]

although the general form of $S^*(r)$ is unaltered. Thus an estimate of 2%
for the Norse component in the Icelandic population is the best that can be
obtained with these data and this series of models, although no value less
than 20% may be rejected.

It is sometimes thought that it is only with respect to the ABO
system that the Icelanders resemble the Irish. We note that the ABO
system does not dominate the distances above. Although it is a significant
contributor, and does provide a major part of d^2_{nc}, it in fact gives

$d_{nc} > d_{in} > d_{ic}$, and hence $\hat{r} > 0$ (Thompson (1973b: Table 1)). Again there are unfortunately insufficient data to examine the consistency of results given by different data sets. Although most of the possible divisions of the six available dimensions into two three-dimensional groups do confirm the estimate of a Norse component of less than 20%, no conclusions can be drawn.

For comparison we consider briefly the sampling model (6.2.3) with quadratic MRS (6.3.12). (6.2.4) gives $\hat{r} = 0.066$, a rather larger Norse component of approximately 7%. From the form of the quadratic function of (6.3.7) we may deduce that this is the largest estimate obtainable with these data and this series of models under the reasonable assumptions that $\sigma_c^2 \geq 0.071\sigma_n^2$ and $6.7\sigma_c^2 + 5.6\sigma_i^2 \geq \sigma_n^2$ the latter condition ensuring form (i) of S^* and the former $g(0.066) \leq 0$. Similarly we have the smallest estimate, $\hat{r} = 0$, if $\sigma_c^2 \geq 0.77\sigma_i^2$ and $6.7\sigma_c^2 + 5.6\sigma_i^2 \geq \sigma_n^2$. The sample size is a scale factor in the MRS (6.3.12), and the support function given by the same genetic distances and a hypothetical Icelandic sample size of 250 is also plotted in Fig. 6.4. It may be seen that $m_i = 250$, or $1/8m_i = 5 \times 10^{-4}$, gives similar uncertainty to the general drift and sampling model with $\sigma_n^2 = \sigma_c^2 = \sigma_i^2 = 2 \times 10^{-4}$, the latter having larger support limits than those given by a quadratic approximation at the maximum. The sampling model gives two-unit support limits

$$0.066 - [15.7/m_i]^{\frac{1}{2}} \leq r \leq 0.066 + [15.7/m_i]^{\frac{1}{2}}$$

for these genetic data. Although, for equal m_i, the sampling model provides far stricter limits for the estimate of r than does the general model, a sampling model is not justified (cf. 1.3). Even with the currently available samples, the distance $\|\underline{h}(\hat{r})\|$ is too large to be explained only by sampling in the Icelandic population.

We have assumed throughout that the variances have been independently estimated. It is however possible to estimate them jointly with the other parameters from the genetic data, by maximising the support (6.3.1) also w.r.t. the three variances. Since a positive lower bound to the variances is provided by the sample sizes we do not have a singularity. In general the ML equations then obtained must be solved iteratively, but provided the relative magnitudes of the variances are known we may obtain an explicit expression. In the case where $\sigma_n^2 = \sigma_c^2 = \sigma_i^2 = \sigma^2$, (6.3.10) may be maximised w.r.t. σ^2 and we obtain the joint ML estimate

$$\hat{\sigma}^2 = h^2(\hat{f})/3pf^*(\hat{f}). \tag{6.4.1}$$

This can provide some check on the model, although it is very far from being a complete test. For had differential selection acted the mean distances between populations would be larger and estimates of σ^2 inflated. For the present data we have $\hat{\sigma}^2 = 1.3 \times 10^{-4}$, which gives remarkable agreement with the estimates of 1 or 2×10^{-4} previously obtained on the basis of population and sample sizes.

The data on which the above inferences are based are limited, but an enlarged set of data is provided by Bjarnason et al. (1973). From the gene frequencies for eleven loci, giving $p = 13$, we obtain

$$d_{ic}^2 = 0.0376, \quad d_{in}^2 = 0.0517 \text{ and } d_{nc}^2 = 0.0406.$$

Again we obtain the estimate $\hat{f} = 0$ and, although d_{ic} is no longer significantly the smallest distance, these data do seem to confirm that the Icelanders were predominantly Celtic. Although the actual maximising point is now $r = -0.71$, $S^*(-0.71)-S^*(0)$ is small, and the two-unit support limit within $0 \le r \le 1$ $[S^*(0)-S^*(r) \le 2]$ is similar to before. The estimate (6.4.1) of σ^2 given by the genetic data (and $\hat{f} = 0$) is now 4.8×10^{-4} This larger value may be accounted for by dominance and the smaller sample sizes on which the data for some loci are based, but may also be an indication of non-representative sampling (6.1), or of selection at some loci.

Finally we consider an alternative hypothesis that has been suggested in order to preserve a predominantly Scandinavian origin for the Icelanders. This is the hypothesis of migration to Norway since A.D. 1000 resulting in the present Norwegians not being descended from the Norsemen of that date. This hypothesis was first mentioned by Donegani et al. (1950), but there is little supporting historical evidence, and while it explains the dissimilarity between present day Icelanders and Norwegians it does not explain the similarity of Irish and Icelanders. The hypothesis thus further requires that the Norsemen of A.D. 900 had Celtic-type gene frequencies and/or that a Norse component also dominates the ancestry of Ireland and Northern Scotland. This does not seem to be tenable. The Norse influence in Ireland and North-Western Scotland was superficial and transitory, and these peoples have always been regarded as of Celtic origin. Further, although the evidence is far from conclusive, in those areas where the Norse influence was more lasting, Orkney and the Isle of Man, the gene frequencies seem to be more similar to those of present day Norwegians (Mitchell (1973), Boyce et al. (1973)). Thus at the present time it seems that the best explanation of the data is a Celtic component in the Icelandic colonising population much larger than has previously been suspected.

Summary

To summarise, we restate the aim proposed in the Preface; that of providing a specific answer to a specific question. On the basis of the likelihood theory of inference, and of a model of bifurcating populations and genetic evolution due to random genetic drift, we have presented an analysis of the inferences of evolutionary history that can be made from current genetic data. In particular, we have produced and studied a procedure for obtaining the maximum likelihood estimate of evolutionary history. Chapter 6 shows that the same general approach may be used to analyse an independent but related problem, although the details of the model, as of the hypotheses to be considered, are different.

Many criticisms may be made of the model, but it should be borne in mind that in likelihood inference we are interested in the relative abilities of hypotheses to describe the way in which current data have arisen: we do not assume that any one hypothesis provides a complete explanation. A model including all the factors that contribute to current genetic data might be preferable in theory, but a model which does not permit the analysis of data is not of practical use. It is clear that, with current genetic data, little could be gained by a more sophisticated evolutionary model; the effect of sampling errors and non-representative sampling will be far greater than those of minor non-isolation or selection, or even of differing population size.

The most serious criticism of the analysis is that, although the effect of sampling and differing population size can be estimated, they are not explicitly included in the model. The inclusion of sampling is feasible in some cases (5.3, Chapter 6), but more generally the alternative must be to obtain genetic data for which the sampling errors are truly negligible. Situations of dominance in sampling and the more fundamental problem of obtaining samples representative of the relevant unit of population (2.2) are outside the scope of the model. We see in Chapter 6 that, for evolution over comparatively short periods of time, sampling may contribute a major part of observed population distances.

However, the results of Chapter 5, together with our knowledge of evolutionary history obtained from other sources, show that even current sample gene frequencies do contain information concerning evolution, and that our model is a sufficiently accurate description of reality to allow some of this information to be extracted, and useful inferences made.

References

Adke, S. R. and Moyal, J. E. (1963). A birth, death and diffusion process. J. Math. Anal. Appl. 7, 209-24.

Barndorff-Nielsen, O. (1971). On conditional statistical inference. Matematisk Institut Publications, Aarhus University.

Bhattacharyya, A. (1946). On a measure of divergence between two multinomial populations. Sankhyā, 7, 401-6.

Bjarnason, O., Bjarnason, V., Edwards, J. H., Fredriksson, S., Magnusson, M., Mourant, A. E. and Tills, D. (1973). The blood groups of the Icelanders. Ann. hum. Genet. 36, 425-55.

Bodmer, W. F. and Cavalli-Sforza, L. L. (1968). A migration matrix model for the study of random genetic drift. Genetics, 59, 565-92.

Boyce, A. J., Brothwell, D. R. and Holdsworth, V. M. L. (1973). Demographic and genetic studies in the Orkney Islands. In: Genetic Variation in Britain, eds. D. F. Roberts and E. Sunderland. London: Taylor and Francis.

Cavalli-Sforza, L. L. and Bodmer, W. F. (1971). The genetics of human populations. San Francisco: W. H. Freeman and Co.

Cavalli-Sforza, L. L. and Edwards, A. W. F. (1964). Analysis of human evolution. In: Genetics Today, ed. S. J. Geerts. (Proc. XI Int. Congr. Genet., The Hague (1963)). New York: Pergamon Press.

Cavalli-Sforza, L. L. and Edwards, A. W. F. (1967). Phylogenetic analysis; models and estimation procedures. Am. J. hum. Genet. 19, 233-57.

Constandse-Westermann, T. S. (1972). Genetical affinities between populations in Western Europe and Scandinavia. In: The assessment of population affinities in man, eds. J. S. Weiner and J. Huizinger. Oxford: Clarendon Press.

Dayhoff, M. O. (1969). Computer analysis of protein evolution. Sci. Am. 221, 86-95.

Donegani, J. A., Dugal, N., Ikin, E. W. and Mourant, A. E. (1950).
The blood groups of the Icelanders. Ann. Eugen. 15, 147-52.

Edwards, A. W. F. (1966). Minimum evolution: program specification.
Unpublished circulated note; University of Aberdeen.

Edwards, A. W. F. (1970). Estimation of the branch points of a branching
diffusion process. J. Roy. Statist. Soc. B, 32, 154-74.

Edwards, A. W. F. (1971). Distances between populations on the basis
of gene frequencies. Biometrics, 27, 873-81.

Edwards, A. W. F. (1972). Likelihood. Cambridge: Cambridge University Press.

Edwards, A. W. F. and Cavalli-Sforza, L. L. (1963). The reconstruction
of evolution. Paper read at 142nd meeting of the Genetical Society,
London. Abstracts in Ann. hum. Genet. 27, 104-5 and Heredity,
18, 553.

Elston, R. C. (1971). Estimation of admixture in racial hybrids. Ann.
hum. Genet. 35, 9-17.

Ewens, W. J. (1965). The adequacy of the diffusion approximation to
certain distributions in genetics. Biometrics, 21, 386-94.

Felsenstein, J. (1968). Statistical inference and the estimation of phylogenies.
Unpublished Ph. D. thesis; University of Chicago.

Felsenstein, J. (1973). Maximum likelihood estimation of evolutionary
trees from continuous characters. Am. J. hum. Genet. 25,
471-92.

Fisher, R. A. (1956). Statistical methods and scientific inference.
Edinburgh: Oliver and Boyd.

Fitch, W. M. and Margoliash, E. (1967). Construction of evolutionary
trees. Science, 155, 279-84.

Fitch, W. M. and Neel, J. V. (1969). The phylogenetic relationships of
some Indian tribes of Central and South America. Am. J. hum.
Genet. 21, 384-97.

Friedlaender, J. S., Sgaramella-Zonta, L. A., Kidd, K. K., Lai, L.Y.C.,
Clark, P. and Walsh, R. J. (1971). Biological divergences in
South-Central Bougainville: an analysis of blood polymorphism
and anthropometric measurements utilising tree models, and a

comparison of these variables with linguistic, geographic and migrational 'distances'. Am. J. hum. Genet. 23, 253-70.

Glass, B. and Li, C. C. (1953). The dynamics of racial intermixture - an analysis based upon the American Negro. Am. J. hum. Genet. 5, 1-20.

Gomberg, D. (1966). 'Bayesian' post-diction in an evolution process. Unpublished; University of Pavia.

Goodman, M., Barnabas, J., Matsuda, G. and Moore, G. W. (1971). Molecular evolution in the descent of man. Nature, 233, 604-13.

Harding, E. F. (1971). The probabilities of rooted tree shapes generated by random bifurcation. Adv. Appl. Prob. 3, 44-77.

Jardine, N. and Sibson, R. (1971). Mathematical taxonomy. London: Wiley.

Jeffreys, H. (1938). Maximum likelihood, inverse probability and the method of moments. Ann. Eugen. 8, 146-51.

Kalbfleisch, J. D. and Sprott, D. A. (1970). Application of likelihood methods to models involving a large number of parameters. J. Roy. statist. Soc. B, 32, 175-208.

Kidd, K. K. and Sgaramella-Zonta, L. A. (1971). Phylogenetic analysis: concepts and methods. Am. J. hum. Genet. 23, 235-52.

Kimura, M. (1955). Solution of a process of random genetic drift with a continuous model. Proc. Nat. Acad. Sci. USA, 41, 144-50.

Kimura, M. (1964). Diffusion models in population genetics. J. Appl. Prob. 1, 177-232.

Kimura, M. and Ohta, T. (1971). Theoretical aspects of population genetics. New Jersey: Princeton University Press.

Kimura, M. and Weiss, G. H. (1964). The stepping-stone model of population structure and the decrease of genetic correlation with distance. Genetics, 61, 763-71.

Krieger, H., Morton, N. E., Mi, M. P., Azevedo, E., Freire-Maia, A. and Yasuda, N. (1965). Racial admixture in North-Eastern Brazil. Ann. hum. Genet. 29, 113-26.

Malecot, G. (1959). Les modèles stochastiques et génétique de population. Pub. Inst. Stat. Univ. de Paris, 8, 173-210.

Malyutov, M. B. , Passekov, V. P. and Rychkov, Y. G. (1972). On the reconstruction of evolutionary trees of human populations resulting from random genetic drift. In: The assessment of population affinities in man, eds. J. S. Weiner and J. Huizinger. Oxford: Clarendon Press.

Maruyama, T. (1972). The rate of decay of genetic variability in a geographically structured finite population. Math. Biosci. 14, 325-35.

Mitchell, R. J. (1973). Genetic factors in the Isle of Man and neighbouring areas. In: Genetic variation in Britain, eds. D. F. Roberts and E. Sunderland. London: Taylor and Francis.

Morton, N. E. , Yasuda, N. , Miki, C. and Yee, S. (1968). Bioassay of population structure and isolation by distance. Am. J. hum. Genet. 20, 411-19.

Morton, N. E. , Yee, S. , Harris, D. E. and Lew, W. (1971). Bioassay of kinship. Theor. Pop. Biol. 2, 507-24.

Neel, J. V. and Ward, R. H. (1970). Village and tribal genetic distances among American Indians, and the possible implications for human evolution. Proc. Nat. Acad. Sci. USA, 65, 323-30.

Post, R. H. , Neel, J. V. and Schull, W. J. (1968). Tabulations of phenotype and gene frequencies for eleven different genetic systems studied in the American Indian. In: Biomedical challenges presented by the American Indian, ed. J. V. Neel. New York: PAHA Scientific Publications.

Thompson, E. A. (1972). The likelihood for multinomial proportions under stereographic projection. Biometrics 28, 618-20.

Thompson, E. A. (1973a). The method of minimum evolution. Ann. hum. Genet. 36, 333-40.

Thompson, E. A. (1973b). The Icelandic admixture problem. Ann. hum. Genet. 37, 69-80.

Thompson, E. A. (1974). Mathematical analysis of human evolution and population structure. Thesis submitted for Ph. D. degree; University of Cambridge.

Ward, R. H. and Neel, J. V. (1970). Gene frequencies and micro-differentiation amongst the Makiritare Indians, IV: A comparison of genetic network with ethnohistory and migration matrices: A new index of genetic isolation. <u>Am. J. hum. Genet.</u> 22, 538-61.

Wright, S. (1943). Isolation by distance. Genetics 28, 114-38.

Wright, S. (1951). The genetical structure of populations. <u>Ann. Eugen.</u> 15, 323-54.

References index

Adke and Moyal (1963), sections 3. 1, 3. 2, 3. 4

Barndorff-Nielsen (1971), 1. 3, 5. 5

Bhattacharyya (1946), 1. 2

Bjarnason et al. (1973), 5. 1, 6. 1, 6. 4

Bodmer and Cavalli-Sforza (1968), 1. 1

Boyce et al. (1973), 6. 4

Cavalli-Sforza and Bodmer (1971), 1. 1, 2. 1, 2. 2

Cavalli-Sforza and Edwards (1964), 1. 2, 1. 4

Cavalli-Sforza and Edwards (1967), 1. 1, 1. 2, 1. 4, 2. 1, 2. 2, 2. 3, 3. 2,
    4. 6

Constandse-Westermann (1972), 6. 1

Dayhoff (1969), 1. 1

Donegani et al. (1950), 6. 1, 6. 4

Edwards (1966), 1. 4, 5. 1

Edwards (1970), 1. 2, 2. 1, 3. 2, 3. 4

Edwards (1971), 1. 2, 2. 3

Edwards (1972), 1. 3, 2. 4

Edwards and Cavalli-Sforza (1963), 1. 2, 1. 4

Elston (1971), 6. 2

Ewens (1965), 2. 1

Felsenstein (1968), 1. 2, 3. 1, 5. 5, 5. 6

Felsenstein (1973), 1. 2, 1. 3, 1. 4, 3. 1, 3. 2, 4. 5, 5. 5

Fisher (1956), 1. 3, 5. 4

Fitch and Margoliash (1967), 1. 1

Fitch and Neel (1969), 1. 2

Friedlaender et al. (1971), 1. 2

Glass and Li (1953), 6. 2

Gomberg (1966), 1. 2, 3. 1

Goodman et al. (1971), 1. 1

Harding (1971), 2.1, 3.1, 3.2, 3.4

Jardine and Sibson (1971), 1.1

Jeffreys (1938), 1.3

Kalbfleisch and Sprott (1970), 1.3, 5.5

Kidd and Sgaramella-Zonta (1971), 1.4

Kimura (1955), 2.3

Kimura (1964), 2.1

Kimura and Ohta (1971), 2.1

Kimura and Weiss (1964), 1.1

Krieger et al. (1965), 6.2

Malecot (1959), 1.1

Malyutov et al. (1972), 1.4, 2.2, 4.6

Maruyama (1972), 1.1

Mitchell (1973), 6.4

Morton et al. (1968), 1.1

Morton et al. (1971), 1.1

Neel and Ward (1970), 2.4

Post et al. (1968), 2.4

Thompson (1972), 2.3

Thompson (1973a), 1.4, 3.3, 5.1, 5.6

Thompson (1973b), 6.1, 6.2, 6.4

Thompson (1974), 1.1

Ward and Neel (1970), 1.2

Wright (1943), 1.1

Wright (1951), 1.1

Subject index

For EU product safety concerns, contact us at Calle de José Abascal, 56–1°,
28003 Madrid, Spain or eugpsr@cambridge.org.